International Symposium on Banda Sea Ecosystem (ISBSE 2017)

AA006071

IOP Conference Series: Earth and Environmental Science
Volume 184

Jakarta, Indonesia
23 October 2017

Editors:

Yosmina Tapilatu
Augy Syahailatua
Zainal Arifin
R. Dwi Susanto

ISBN: 978-1-5108-7010-9
ISSN: 1755-1307

Printed from e-media with permission by:

Curran Associates, Inc.
57 Morehouse Lane
Red Hook, NY 12571

Some format issues inherent in the e-media version may also appear in this print version.

Copyright© (2017) by the Institute of Physics
All rights reserved. The material featured in this book is subject to IOP copyright protection, unless otherwise indicated.

Printed by Curran Associates, Inc. (2018)

For permission requests, please contact the Institute of Physics at the address below.

Institute of Physics
Dirac House, Temple Back
Bristol BS1 6BE UK

Phone: 44 1 17 929 7481
Fax: 44 1 17 920 0979

techtracking@iop.org

Additional copies of this publication are available from:

Curran Associates, Inc.
57 Morehouse Lane
Red Hook, NY 12571 USA
Phone: 845-758-0400
Fax: 845-758-2633
Email: curran@proceedings.com
Web: www.proceedings.com

Table of contents

Volume 184

International Symposium on Banda Sea Ecosystem (ISBSE) 2017

23 October 2017, Balai Kartini Convention Center, Jakarta, Indonesia

Accepted papers received: 1 August 2018
Published online: 20 August 2018

Preface

Papers

Review Papers

Original Research Papers

IOP Conf. Series: Earth and Environmental Science **184** (2018) 011001 doi:10.1088/1755-1315/184/1/011001

Foreword

The Banda Sea was the object of research activities since the 20th century, starting from expeditions carried out during the 1950s. One of the most prominent expeditions ever accomplished in this water was the Snellius Expedition. A joint collaboration between research scientists from Dutch NIOS and the Indonesian Institute of Science (LIPI) Ambon Office, this expedition resulted in the publication of three volumes of substantial findings. These findings have helped to shape the knowledge and information of this tropical deep-sea. LIPI scientists published an oceanology atlas in 1999, but not much afterward was done to compile and critically review what has been done fifteen years after.

The Center for Deep-Sea Research (PPLD), Indonesian Institute of Sciences (LIPI) has held successfully an International Symposium on Banda Sea Ecosystem (ISBSE) 2017 in Balai Kartini Jakarta on October 23rd, 2017. This International Symposium -the first about this particular ecosystem- was designed to carefully review what has been done in the past sixty years, their scientific contribution to the current understanding of this water, and what perspective gained from these research activities for the future management and conservation of its resources. Designed to be a one-day plenary lecture, this event was included in the Conference Series as one of the main activities of Indonesia Science Expo (ISE 2017), within the celebration of LIPI 50th establishment entitled LIPI contribution in building the nation.

Other than LIPI's research institutes (PPLD, Research Center for Geotechnology, LPKSDMO Pari, Research Center for Oceanography, and LKBL Bitung), invited speakers, registered poster presenters and audience came from a varied background of international and national levels. International speakers and audience were from universities and government institution in the USA (University of Maryland), Australia (Australian National University, University of New South Wales and Department of Fisheries of the Western Australia Government) as well as a French consultant company (CreOcean Indonesian office). At the national level, speakers were invited from Conservation International-Indonesian Office, Gadjah Mada University, Research and Development Centre of Marine Geology (P3GL), and Directorate General of Marine Spatial Management (DG MSM) Indonesian Ministry of Marine Affairs and Fisheries (KKP). Registered poster presenters and audiences were mainly lecturers and undergraduate students from Gadjah Mada University, University of Bengkulu, Bogor Agricultural University (IPB), and Satya Wacana Christian University (UKSW).

This symposium has a strategic value as the first international symposium held by PPLD since its inauguration in 2014. It was designed not only as an occasion for exchanging scientific findings and future perspectives regarding research activities on Banda Sea ecosystem among interested stakeholders but also as a starting point towards long-lasting collaboration, leading to the advancement of knowledge and science about the marine environment, notably in the eastern area of Indonesia.

Limitations in funds, infrastructure, and human resources did not hinder the symposium preparation and organization, and this was proven by the publication of this conference proceeding. Rigorous peer-review process has resulted in the publication of twelve selected articles, consisting of seven reviews and five original research papers mostly focused on the marine biology of the Banda Sea, as well as a glance on its geology, chemistry, physic and pollution aspects.

I wish to thank the contributing authors who have submitted the papers and all the reviewers who have strived to ensure that this proceeding meets the high scientific standard required. I am also grateful to IOP Publishing, especially Mrs. Stephanie Gill (IOP Conference Publishing Coordinator), for allowing us to share some of the symposium results with a larger audience of the scientific community through the IOP Conference Series: Earth and Environmental Science. Thanks are also due to the 2017 ISE Chair of Organizing Committee, Dr. L.T. Handoko who has allowed the symposium to be part of the ISE event. I wish as well to acknowledge Dr. Zainal Arifin, the Deputy of Earth Science LIPI who has morally and intellectually supported the symposium and the proceeding conception. I am also indebted to Dr. Augy Syahailatua, Director of PPLD LIPI, who has given his endless support from day one of symposium conception to proceeding publication through fruitful discussions on multiple occasions. This symposium and proceeding are financially supported by PPLD LIPI DIPA funding for the fiscal year 2017 and 2018.

Content from this work may be used under the terms of the Creative Commons Attribution 3.0 licence. Any further distribution of this work must maintain attribution to the author(s) and the title of the work, journal citation and DOI.
Published under licence by IOP Publishing Ltd

IOP Conf. Series: Earth and Environmental Science **184** (2018) 011001 doi:10.1088/1755-1315/184/1/011001

Yosmina Tapilatu
Editor-in-Chief
Proceeding of ISBSE 2017

Conference Chair
International Symposium on Banda Sea Ecosystem 2017

IOP Conf. Series: Earth and Environmental Science **184** (2018) 011002 doi:10.1088/1755-1315/184/1/011002

Editorial Committee

Yosmina Tapilatu, *Center for Deep-Sea Research, Indonesian Institute of Sciences (PPLD LIPI), Ambon 97233, Indonesia*

Augy Syahailatua, *Center for Deep-Sea Research, Indonesian Institute of Sciences (PPLD LIPI), Ambon 97233, Indonesia*

Zainal Arifin, *Research Center of Oceanography, Indonesian Institute of Sciences (P2O LIPI), Jakarta 14430, Indonesia*

R. Dwi Susanto, *Department of Atmospheric and Oceanic Science, University of Maryland, Maryland 20742, USA*

Scientific Committee

Agung Dhamar Syakti, *Raja Ali Haji Maritime University, Indonesia*
Agus Santoso, *University of New South Wales (UNSW), Australia*
Andreas Kunzmann, *Leibniz Centre for Tropical Marine Research (ZMT), Germany*
Augy Syahailatua, *Indonesian Institute of Sciences (LIPI), Indonesia*
Charles PH Simanjuntak, *Bogor Agricultural University (IPB), Indonesia*
Iin Inayat Al Hakim, *Indonesian Institute of Sciences (LIPI), Indonesia*
Jonathan M. Pownall, *Australian National University, Australia*
Novik Nurhidayat, *Indonesian Institute of Sciences (LIPI), Indonesia*
Nugroho D. Hananto, *Indonesian Institute of Sciences (LIPI), Indonesia*
R. Dwi Susanto, *University of Maryland, USA*
Roni Bawole, *University of Papua, Indonesia*
Sem Likumahua, *Indonesian Institute of Sciences (LIPI), Indonesia*
Suyadi, *Indonesian Institute of Sciences (LIPI), Indonesia*
Widya N. Satrioadjie, *Indonesian Institute of Sciences (LIPI), Indonesia*
Yosmina Tapilatu, *Indonesian Institute of Sciences (LIPI), Indonesia*
Zainal Arifin, *Indonesian Institute of Sciences (LIPI), Indonesia*

Content from this work may be used under the terms of the Creative Commons Attribution 3.0 licence. Any further distribution of this work must maintain attribution to the author(s) and the title of the work, journal citation and DOI.
Published under licence by IOP Publishing Ltd

IOP Conf. Series: Earth and Environmental Science **184** (2018) 011003 doi:10.1088/1755-1315/184/1/011003

ISBSE 2017

International Symposium on Banda Sea Ecosystem

Pusat Penelitian Laut Dalam
Center for Deep-Sea Research
LIPI

October 23rd 2017, Jakarta

INDONESIA SCIENCE EXPO

Content from this work may be used under the terms of the Creative Commons Attribution 3.0 licence. Any further distribution of this work must maintain attribution to the author(s) and the title of the work, journal citation and DOI.
Published under licence by IOP Publishing Ltd

IOP Conf. Series: Earth and Environmental Science **184** (2018) 011004 doi:10.1088/1755-1315/184/1/011004

Peer review statement

All papers published in this volume of *IOP Conference Series: Earth and Environmental Science* have been peer reviewed through processes administered by the proceedings Editors. Reviews were conducted by expert referees to the professional and scientific standards expected of a proceedings journal published by IOP Publishing.

Content from this work may be used under the terms of the Creative Commons Attribution 3.0 licence. Any further distribution of this work must maintain attribution to the author(s) and the title of the work, journal citation and DOI.
Published under licence by IOP Publishing Ltd

Banda deep-sea research: history, mission and strategic plan

Suyadi[1,2], W N Satrioajie[1], A Syahailatua[1] and Z Arifin[3]

[1]Center for Deep-Sea Research, Indonesian Institute of Sciences (PPLD LIPI), Jl. Y. Syaranamual Guru-Guru Poka Ambon 97233, Indonesia
[2]School of Environment, University of Auckland, New Zealand
[3]Research Centre for Oceanography - Indonesian Institute of Sciences (P2O LIPI), Indonesia

suya009@lipi.go.id, yadi_pdt@yahoo.com

Abstract. Attention to the deep-sea ecosystem research in order to achieve blue economy goals is low. The Banda Sea is important deep-sea ecosystem in Indonesia and it was identified as a global marine mega-biodiversity site. The objectives of this paper are 1) to document the history, achievements, and challenges of deep-sea research development and 2) to develop a grand design and innovative strategic plan for deep-sea research. Literature reviews and questionnaire survey for scientists across Indonesia were carried out. The literature reviews indicated that deep-sea ecosystem is not well documented. Few articles on deep-sea were published (mean 0.8 article y^{-1}) which were mainly focused on geological (45%) and physical oceanography (32%), less research on biological (16%) and chemical oceanography (8%). Deep-sea ecosystem faced tremendous threats due to ecosystem degradation, pollution, and climate change. The questionnaire survey showed that research infrastructure/facilities (Ranking 1) and institutional management systems (Ranking 2) were the main obstacles to Indonesian deep-sea research development. A grand design named Indonesian Deep-sea Research and Monitoring (I-DREAM) and its strategic plan were developed to achieve blue economy goals. Strengthening research infrastructure and institutional management systems are the top priories that should be achieved in the short-term period to support deep-sea research in Indonesia. Scientific pursuit in long-term period should be focused on biodiversity and deep-sea ecology including chemical oceanography (24%) and its interaction with other ecosystems and atmosphere (18%) were identified as the top priorities to be implemented for future research (Ranking 1 and 2). The Banda Sea was identified as a suitable core study area of the I-DREAM's pilot project. The grand-design and its innovative strategic plan are essential to fulfilling high demand on a guideline for deep-sea research to achieve blue economic development goals.

Keywords: Deep-sea, blue economy, grand design, strategic plan, Banda Sea, marine research

1. Introduction

Around 75% of Indonesian territory is sea and most of Indonesia seas have deep-sea zones (below 200 m) [1]. Located in the tropical region, making the Indonesian seas biologically rich resources and rich of energy resource such as oil, gas, and minerals [1,2]. The Banda Sea (± 0.54 million km^2) is the largest deep-sea in Indonesia located in Maluku Province [3]. In addition, the Banda Sea is known as one of the deepest seas (7440 m) in the world [4,5].

Content from this work may be used under the terms of the Creative Commons Attribution 3.0 licence. Any further distribution of this work must maintain attribution to the author(s) and the title of the work, journal citation and DOI.
Published under licence by IOP Publishing Ltd

The Banda Sea is a marine mega-biodiversity centre situated in the Coral Reef Triangle Initiative area. Marine biota and various ecosystems such as mangroves, seagrass, and coral reef are found in this seascape [6, 7]. In 2009, the ministry of Marine Affairs and Fisheries (Ministerial Decree No. 60 2009 and No. 58 2014) declared the Banda Sea as a national marine protected area of 25 km^2 [8]. It is including Banda Islands, located in the middle of the Banda Sea (Figure 1).

It is predicted that human population in Indonesia will grow up to 300 million in 2035 [9] and Indonesia will face both biological and natural resources crisis especially food and energy [10-12]. Deep-sea ecosystems provide ecological functions and ecosystem services and rich of biological resources [13-15]. These potential of deep-sea are important to achieve blue economy goals especially to support food and energy security and marine infrastructure [14]. For example, marine biota could be used as a raw material of medicine and food. Deep-sea potentially have alternative energy resources and important for infrastructure development (e.g. communication network, tunnel, etc.).

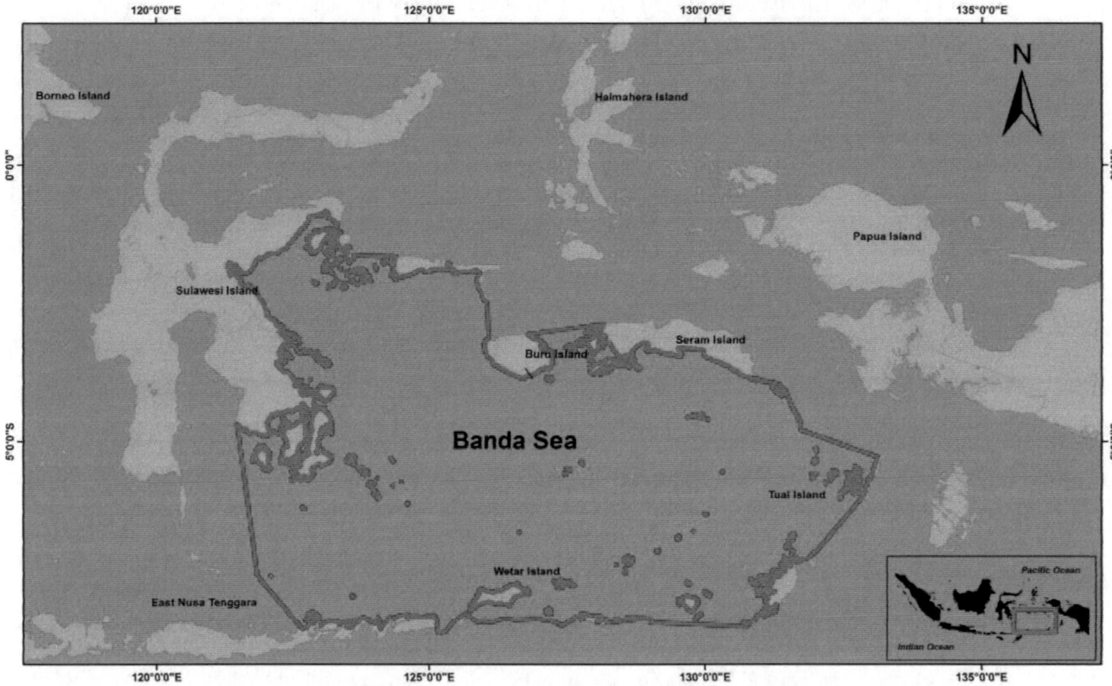

Figure 1. The Banda Sea is located in the eastern part of Indonesia.

The blue economy is a concept of sustainable use of sea resources for economic growth, improved livelihoods and jobs, and sea ecosystem health [16]. The blue economy goals include six main aspects: tourism, renewable energy (i.e., OTEC/Ocean Thermal Energy Conversion, which is a process that can generate electricity by using the temperature difference between deep cold ocean water and warm surface of the Banda Sea water), climate change, fisheries, maritime transport and infrastructure, waste management and conservation [16]. This concept includes deep-sea, small islands, coastal areas, and shallow sea. However, the deep-sea resources and its ecosystem conditions are not well explored. Deep-sea ecosystems face tremendous threats due to human activities and environmental change, such as environmental degradation, climate change, and water pollution [17-19]. In addition, scientific information on the interconnectivity of deep-sea, coastal, and terrestrial ecosystem is lacking.

Deep-sea research and monitoring are highly required to achieve the blue economy goals especially to mitigate the biological and natural resources crisis and to ensure sustainable use of the resources through conservation actions. However, grand-design and strategic plan of deep-sea research are not available yet. These documents are required to ensure that deep-sea research activities are effective to achieve the blue economic goals. This study is the first effort to review Indonesian deep-sea research and to develop a grand design for future deep-sea research. The objectives of this study were: (1) to document the history, achievement, and challenge of deep-sea research in Indonesia including in the Banda Sea; and (2) to develop a grand-design and strategic plan of deep-sea research to achieve blue economy goals. Literature review and questionnaire survey were conducted to address the described objectives.

2. Methodology

Two main methods were applied: literature review and questionnaire survey (Figure 2). A literature search was conducted to identify published literature containing information on deep-sea research in Indonesia from the last three centuries (1872-2017). Challenger expedition (1872) was used as starting point of literature review because this expedition was known as the first milestone for oceanography studies in Indonesia [20]. Relevant peer-reviewed studies within Scopus, ISI Web of Science (Thomson Reuters), and Google Scholars were searched using keywords combination such as "deep-sea" AND "Indonesia" or "deep-sea" AND "Banda Sea". Only original research paper published in international, national and local journals/proceedings and graduate thesis were included in the study. Hand-searching was conducted in Indonesian Institute of Sciences (LIPI) libraries for articles which were not published online. Secondly, articles obtained were filtered manually to check for their relevance to the deep-sea research. Lastly, the obtained articles were categorized according to four main subjects: geological oceanography, physical oceanography, chemical oceanography, and biological oceanography.

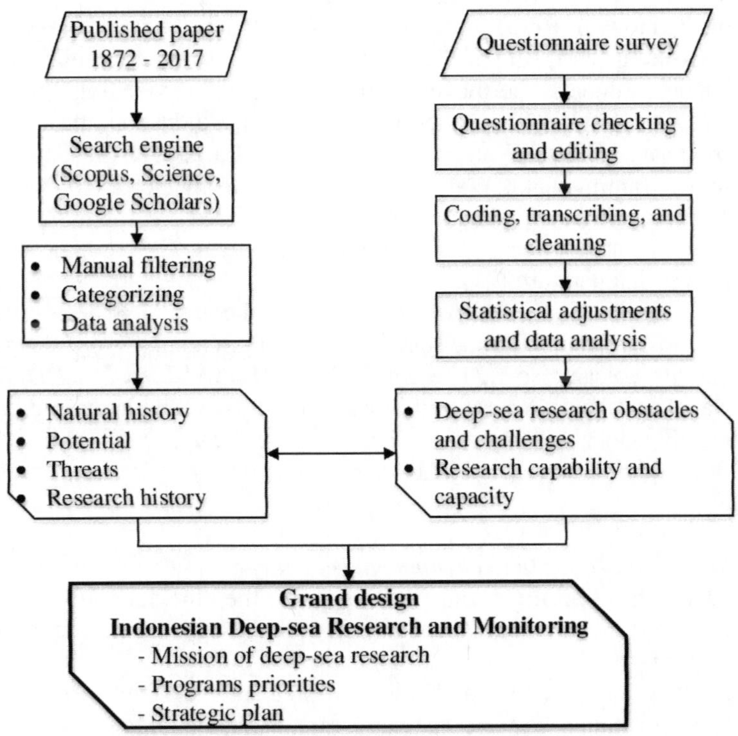

Figure 2. Flowchart of methods: literature review and questionnaire survey used in the study.

A questionnaire survey was conducted in August – October 2017 for 28 respondents in total. The respondents were scientists/researchers (75%) and lecturers (25%) across Indonesia with master and doctoral degree and minimum research experience of eight years. The demography of the respondents was: ages = 29 – 60 years, male = 74%, and female = 26%. There were two types of questions: open-ended questions and close-ended questions (response-option questions). The responses of the questions were designed as single responses, multiple responses, ranked responses and rated responses. There were three mains steps in data processing (figure 1): (1) questionnaire checking and editing to eliminate unacceptable questionnaires and to correct inconsistent and ambiguous answers; (2) coding to assign alpha or numeric codes, transcribing, and cleaning for consistencies; (3) statistical adjustment and data analysis. Statistical adjustment is required to data that requires weighting and scale transformations. The data were analyzed using descriptive analysis in SPSS version 22 (IBM, New York, USA) and the data were normalized using z-score to vary between 0 and 1, thus allowing a comparison and a rank over the different calculated analytical values [21].

3. Result and discussion

3.1. Natural history of the Banda Sea

Located in eastern Indonesia (3.47°S – 7.65°S and 122.42°E – 131.47°E), the Banda Sea (± 0.54 million km^2) is important waters surrounded by small islands of Southeast Sulawesi, Ceram, Wetar, Timor and Tanimbar islands [3]. The Banda Sea consists of six basins within ranges from 2000 m to 7440 m. It has the greatest active arc and active volcanos in Indonesia, making the seascape and islands geologically dynamics [22, 23]. The maximum depth 7440 m was located in the Weber Basin [24]. The depth of the Banda Sea was classified into three zones: bathypelagic (200 - 1000 m), abyssopelagic (1000 - 4000 m), and hadalpelagic zone (>4000 m) [25].

The sea was subjected to two winds: northwest monsoon (from the west, November to March) and southeast monsoons (from the east, May to September) with transitional months on October and April [3, 26]. These monsoons influence the movement of water mass (upwelling and down-welling) and water circulation [27-29]. Upwelling happened on the southeast monsoon and downwelling occurred on the northwest monsoon [3, 30]. Water temperature and salinity in the bathypelagic zone were described as Antarctic Intermediate Water which flows from north to western Pacific. The characteristics of water in the abyssopelagic were not influenced by Antarctic Bottom Water [25]. Water temperature in the abyssopelagic was higher (4.8–5.25 °C) compare to the international standard of abyssopelagic temperature [31].

The Banda Islands, which is situated in the middle of the Banda Sea, consist of inner islands (Banda Naira, Banda Besar, Pisang Island, Karaka Island, and Gunung Api Island) are mountainous and forested islands, with volcanic soils. The outer islands (Hatta Island, Ay Island, Nailakka Island, Rhun Island, and Manukan Island) are uplifted limestone and they are flatter and dryer than the inner islands [32]. Traditionally, local people on the islands have local wisdom called "sasi" to conserve marine resources [34, 35]. However, this local wisdom is under pressure and almost disappeared due to modernization [35].

3.2. Potential of the Banda Sea for blue economy development

The Banda Sea is identified as an important site to support the blue economy development and blue carbon through biodiversity, tourism and transportation, and deep-sea mining as discussed in detail below [36, 37].

Biodiversity centre: the Banda Sea and adjacent waters are habitats for many marine biotas including 133 species of mangroves, 91 species of mollusc, and many species of fish. Maluku including the Banda Sea is the habitat for 128,000 ha of mangrove forest [6]. The Banda Sea contributes about 4.3% of the national fisheries such as tuna and pelagic fish [38]. This area was declared by the Indonesian Government (Kepmen No. 714) as the main fishing ground for pelagic fish and tuna. The other important marine resources are shrimps, lobsters, coral, and demersal fish [39]. In addition, the eleven small islands in the Banda Sea were the world's sole source of nutmeg made the Banda Islands called as the "Spice Islands" [32, 33].

Tourism and transportation: the Banda Sea and the islands are popular tourist destinations for scuba diving, marine conservation, historical heritage, and traditional culture [40]. It was not clear how many local and international tourists visiting the Banda Sea and the islands. However, the Banda Neira Airport tourist information service estimated of 600-1,500 foreign tourists visited the Banda Sea and the islands during 1990-1994. The Banda Sea is an important marine transportation network linked eastern Indonesia (e.g. Ambon, Tual, and Papua) and western Indonesia (Sulawesi, Java, and West Nusa Tenggara, etc.) [41]. In addition, the Banda Sea is a "bridge" connecting northern and southern part of the Maluku Province [42].

Mining: the Banda Sea has good potential for oil, gas, and mineral resources, but it is not explored well [43]. Regardless of the controversy of the mining system inshore or offshore, the exploration provides a new perspective about the deep-sea mining for energy supply to local society. For example, the development Liquid Natural Gas (LNG) mining of Masela Block located in two small islands was started in early 2000 [43].

3.3. Threats of Banda Sea

Generally, the threats of the Banda Sea can be categorized as anthropogenic and natural threats. Environmental degradation (e.g. mangrove deforestation and pollution) and overexploitation (e.g. overfishing, coral-reef exploitation) are main anthropogenic threats. For example, mangrove forest area in Maluku including the Banda Sea decreased from 198,000 ha in 1980 to 100,000 ha in 2005 [6]. Recently, mangrove area in the Ambon Bay, part of the Banda Sea, was decreased by 10.53% at the rate of 1.5% y^{-1} due to coastal area development [44]. Overfishing is a new threat since the increasing of purse seine fisheries operated in the Banda Sea.

The impacts of climate change such as sea level rise and unpredictable weather (heavy rain, storm, and high tide) were tremendous threats to the deep-sea ecosystem [45]. An unprecedented threat such as mortality of large marine mammals was recorded in the Banda Sea due to climate change associated with changing in the monsoons [14, 19, 46]. Household waste and pollution were other anthropogenic threats. Around 60% of Indonesian plastic waste dumped into the sea. [47]. Illegal activities of gold mining in the Buru Island, for example, polluted the waters and marine ecosystem.

The Banda Sea is located in the ring of fire where a large number of earthquakes, volcanic eruptions, and tsunami were recorded. This is due to the Banda Sea was created as a basin and active subduction zone of Sunda tectonic plate and Sahul tectonic plate. The Indonesian Agency for Meteorology, Climatology, and Geophysics recorded 1,222 earthquake events occurred in Maluku in 2016. This natural disaster might cause environmental degradation of the deep-sea ecosystem.

3.4. Research history in the Banda Sea

The literature review showed that deep-sea research activities in the Banda Sea have been started since the 19th century when the Challenger expedition (1872-1876) took place in the Banda Sea [20]. Afterward, several international and local expeditions were undertaken in the Banda Sea including Galathea expedition (1950-1952), SEATAR (Studies on South East Asia Tectonics and Resources) expedition (1973), Snellius II (1984-1985), and the Banda Sea expedition (November 2013). Most of these expeditions focused on geological and physical oceanography. The concept of plates tectonic in the Banda Sea, for example, was founded by the SEATAR expedition [20]. This expedition identified some locations of gas, oil, and offshore mining [20]. The phenomena of Weber Basin was introduced by Galathea expedition [20]. Some marine biota such as sea cucumber (*Paroriza grevei*), polychaeta, isopods (*Macrostylis hadalis*) and (*Lepthanthura hendili*), and numphon (*Numphon femorale*) were collected during the Galathea expedition. The organic productivity of phytoplankton was the major finding of this expedition [20].

There were only 128 original research papers about Indonesian deep-sea were published since the last three centuries (1872-2017). About 9% of the papers are related with the deep-sea ecosystem of the Banda Sea. This means there were only 0.8 articles of Indonesian deep-sea published every year. Specifically, there was only –in average- 1.3 articles about the Banda deep-Sea published every decade. The number of publication about deep-sea in Indonesia is lower than the Philippines (mean 0.9 article y^{-1}) and much lower than the United States of America (4 articles y^{-1}).

The subjects of the published paper varied. In this paper, the research topics were categorized into four categories of subject/areas for the convenience of discussion: 1) geological oceanography (e.g. geophysics, geochemistry, geo-hazard etc.); 2) physical oceanography; 3) chemical oceanography; and 4) biological oceanography (ecology, marine environment, fisheries, microbiology, marine biology. Given to limited articles focusing on integrated topics such as air-sea interaction and climate and ecosystems (deep-sea, coastal, and terrestrial) interaction, these topics included in biological and physical oceanography. The literature review revealed that almost half of the published papers (45%) are related with geological oceanography, followed by physical oceanography (32%). Plate tectonic, geomorphology, and sedimentology of the seafloor are the most common subjects of geological oceanography. The past studies on physical oceanography of deep-sea mainly focused on descriptive physical oceanography (e.g. temperature, salinity, density, etc.) and dynamics of physical oceanography (e.g. water mass motion). Research activities on biological and chemical oceanography are limited, 16% and 8%, respectively.

Generally, previous deep-sea studies described above mainly focused on basic research such as inventory and partial observation. In addition, the deep-sea studies were undertaken limited in several locations such as Sunda/Java trench in the Indian Ocean, Banda trench in Maluku, and Sulawesi basin. This means both basic and advanced research activities of deep-sea especially biological and chemical oceanography are essential. Research on marine biology, biodiversity, deep-sea ecosystem and integrated research topics such as air-sea interaction and climate and ecosystems (deep-sea, coastal, and terrestrial) interaction are required to strengthen conservation actions to ensure availability of marine resources and to mitigate climate change. Advanced research on geological and physical oceanography (e.g. geo-hazard, dynamic structure of deep-sea ecosystems, geologic and climatic prediction, etc.) are still important to support marine transportation and infrastructure development and to develop early warning system of natural disasters such as earthquake and tsunami.

Given the importance of the Banda Sea, in terms of oceanography features, the Banda Sea plays an important role in the establishing of the Indonesian Throughflow (ITF). ITF is considered to have an influence in season phenomena such El Niño Southern Oscillation (ENSO) and Asian-Australian monsoon via global ocean thermohaline circulation and a trade-off on water mass between the Pacific and the Indian Ocean [48]. Indonesian weather is also known to be influenced by oceanographic conditions through air-sea interaction in which the variability of sea surface temperature of the Banda Sea may affect the rainfall [49]. Because of its oceanography dynamics, the upwelling in the Banda Sea supports high diversity of the marine ecosystem. High concentration of chlorophyll-a lead to high productivity in which the Banda Sea has been long known as the potential fishing ground for various fisheries [50-52]. This means that integrated and comprehensive research is crucial not only to support national blue economy goals but also international scales particularly the Pacific and the Indian Ocean.

3.5. Deep-sea research mission and challenges

Maritime is Indonesian national development priority declared in 2014. The mission of deep-sea research in Indonesia is generating excellent scientific based information and technology to support deep-sea management, conservation, and human well-being. Deep-sea research should provide big-significant-real impacts on both human development and effective sustainable management and conservation.

The questionnaire survey showed that the proportion of publication on deep-sea is minimal (2%) with the participation level of respondent on the deep-sea research is only 25%. This finding is supported by literature review that scientific articles related to the deep sea ecosystem are very limited. Similar to literature survey, our questionnaire survey found that physical oceanography, plankton, and geological oceanography are the most common research subjects (82%) of the deep-sea publication.

Proportion and participation of respondents on international dissemination (international scientific publication, conferences, research collaboration) are much lower than local dissemination (table 1). For example, the contribution of respondents in delivering their research findings in international conferences is much lower than in national ones, 27%, and 73%, respectively. Most of the deep-sea research activities were funded by domestic research grants (64%) compare to international ones (36%).

Table 1. Capacity of human resources in publication, dissemination, and research fundraising.

Description	Mean	Proportion (%)	Participation (%)
Publication:			
Publication related to deep sea	0.25	2	25
Publication in international journals	1.38	11	44
Publication in national accredited journals	9.31	72	94
Other publication (popular, newspaper, etc.)	2.19	17	63
Scientist involvement:			
Speaker at international conference	1.75	27	56
Speaker at national conference	4.75	73	81
Funding:			
External (international) research grant	1.00	36	31
Internal (domestic) research grant	1.75	64	56

Research infrastructure (laboratories) and facilities (equipment, research vessel, research funding) are the main obstacles in the development of deep-sea research. Figure 3 illustrates the current conditions of deep-sea research infrastructure and facilities summarises a normalized to vary between 0 (not available) and 1 (good). Firstly, it is pertinent to note that research infrastructure: laboratories, equipment, research vessel, funding for research and publication are in very bad condition (< 0.5, scale 0-1). Secondly, organisation/institutional management systems and collaboration are in bad condition (0.5 – 0.75, scale 0-1). The only good conditions are human resources and literature access. This finding indicates that both current research infrastructure and institutional management system are not ideal to support deep-sea research. Significant improvements in these sectors are required to achieve the blue economic goals through the research related to the deep-sea environment.

The study identified four other challenges in the deep-sea research development associated with managerial/institutional and technical factors. Firstly, the absence of a national grand design (master plan) and strategic plan of deep-sea research resulted in the research priority remained unclear and the research topics and location are often overlapped. Secondly, the institutional and regulation framework of deep-sea research are not effective in the implementation, consequently research collaboration and dissemination do not work properly. Thirdly, lack of standardized methods and long-term replicated measurement in deep-sea research resulted in incompatible data which cannot be used for comparison or upscaling into large scales. Lastly, most of the data and output from deep-sea studies are not managed properly using standardized metadata or were inaccessible.

3.6. Grand design and strategic plan for future deep-sea research

Deep-sea research development is important for Indonesia to support national development priority in order to achieve blue economy goals. Grand design (masterplan) and strategic plan of deep-sea research are essential as a guideline for future deep-sea research. We developed a grand design and its strategic plan named Indonesian Deep-sea Research and Monitoring (I-DREAM) (figure 4). The short-term target or objective (2018-2025) of this program is improving research infrastructure and facilities and strengthening institutional systems. The long-term target or objective (2026-2045) of the program is generating advanced scientific based information on deep-sea ecosystems to support national development priorities: food and energy security, early warning system on the status of biodiversity and ecosystems services, national defense, and disaster mitigation and adaptation.

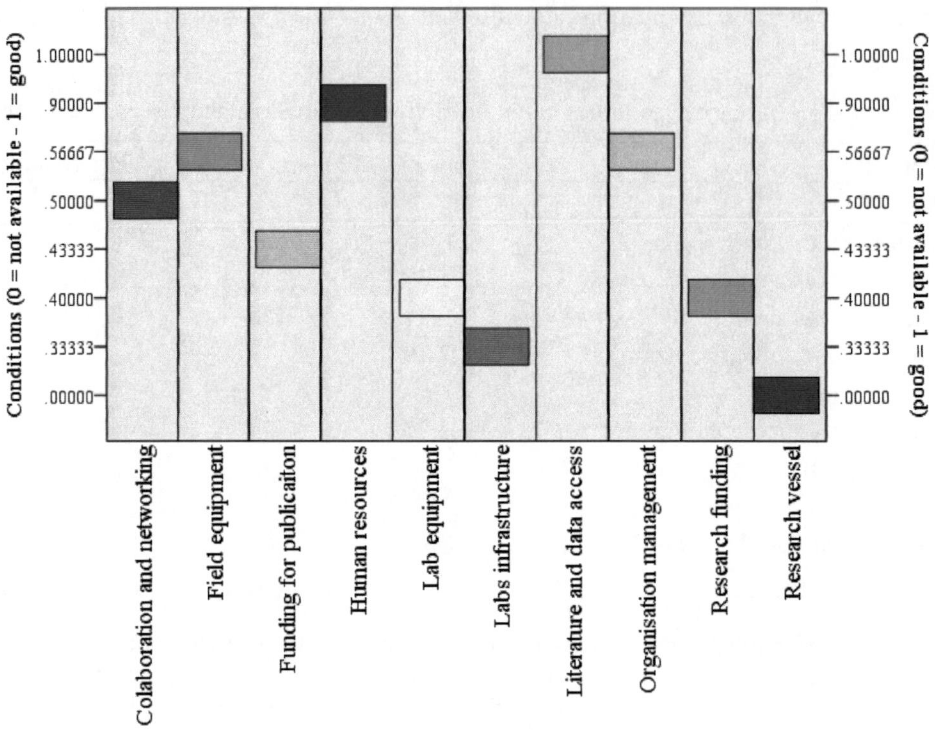

Figure 3. Current condition of deep-sea research infrastructure and facilities.

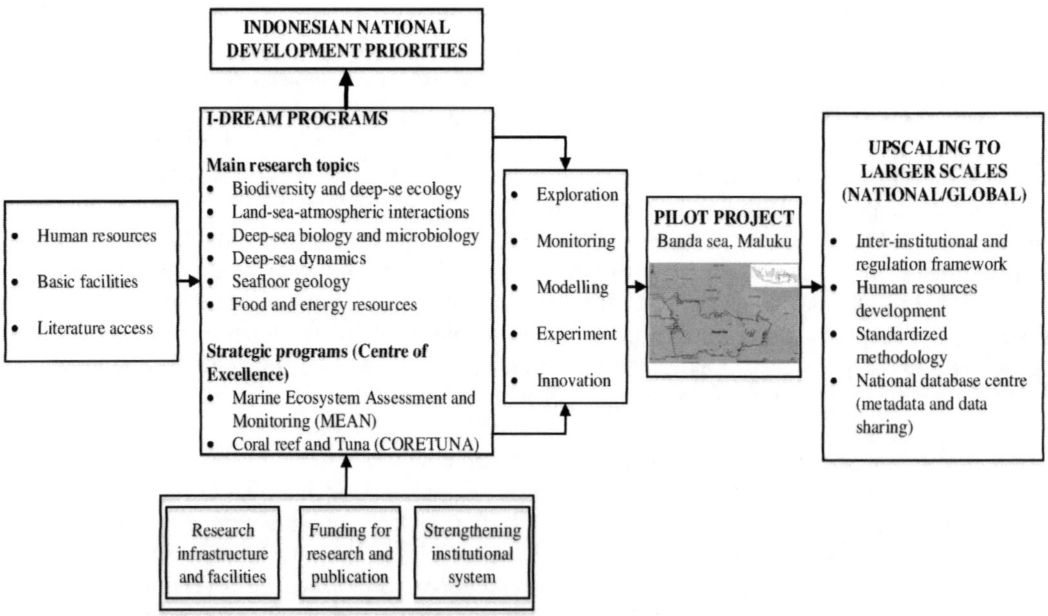

Figure 4. Concept of Indonesian Deep-sea Research and Monitoring (I-DREAM).

The study showed that human resources, access to literature, and basic facilities are available to support deep-sea research in Indonesia. These aspects are basic requirement and important to implement I-DREAM Programs. However, there are three weaknesses should be improved to support the I-DREAM programs. Firstly, improving research infrastructure and facilities such as research vessels and advanced laboratories with adequate equipment for sample collection and analysis are required as research tools. These are essential due to deep-sea zones are untouchable without using equipment (technology). Secondly, sustainable funding for research and publication is crucial to ensure long-term assessment and monitoring of deep-sea ecosystem. Lastly, strengthening institutional systems such as improving management system and efficient bureaucracy are needed for effective research and accountability. The questionnaire survey showed that improving research infrastructure and advanced equipment as the top priority (Ranking 1), followed by strengthening institutional systems and human resources development, ranking 2 and 3, respectively (figure 5a).

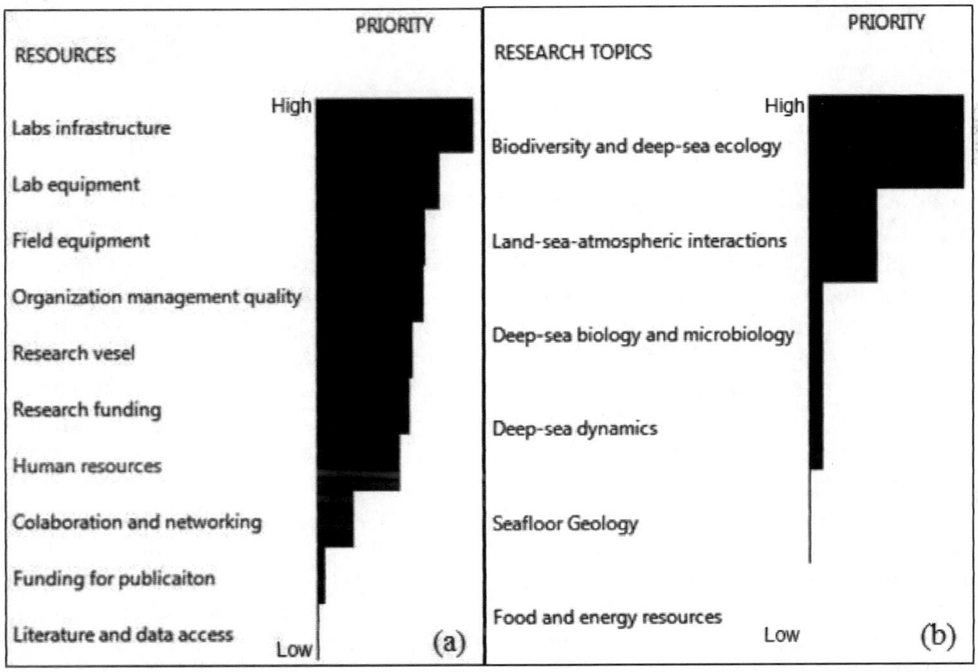

Figure 5. (a) Priority of resources to be improved to support deep-sea research and (b) priority of research topics for deep sea-research.

The questionnaire survey revealed six main research topics for deep-sea research: biodiversity and deep-sea ecology, land-sea-atmospheric interactions (including air-sea interaction and climate), deep-sea biology and microbiology, food and energy resources, deep-sea dynamics, and seafloor geology. The first four topics are associated with biological and chemical oceanography, the fifth topic is physical oceanography, and the last topic is geological oceanography. The equal proportion for the six main research topics and five research categories: exploration, monitoring, experiment, modelling, and innovation are important to provide a holistic and advanced assessment of deep-sea ecosystem (figure 6). The top priority of research topics for future is research related with biological and chemical oceanography (Ranking 1 – 3), followed by deep-sea dynamics (Ranking 4) and seafloor geology (Ranking 5) (Figure 5b). In addition, two strategic programs (figure 4): 1) Marine ecosystem assessment and monitoring (MEAN) and 2) deep-sea coral reef and tuna fisheries (CORETUNA) are suggested as the centre of excellence for deep-sea research. These strategic programs are proposed because they faced tremendous threats from climate change and direct human actions such as overexploitation, pollution, and environmental degradation and required immediate actions to protect them [17, 18]. The centre of excellences is important to improve institutional capacity and capability in specific aspects, which, in turn, accelerate science development and its impacts on human well-being and conservation.

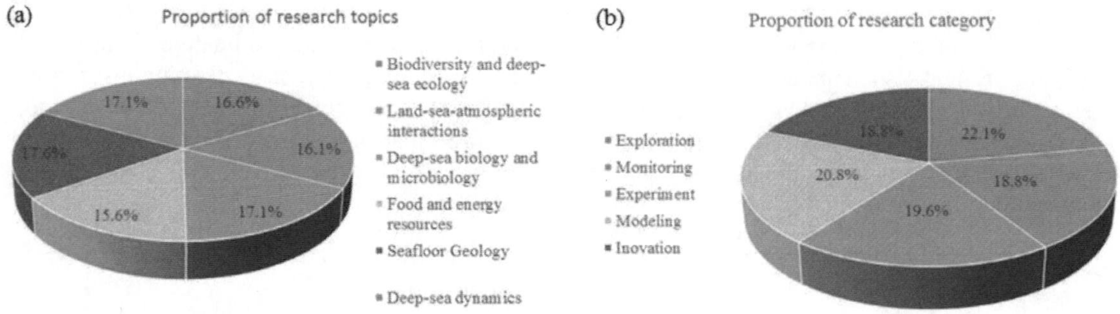

Figure 6. (a) Proportion of dee-sea research topics should be implemented and (b) proportion of research categories for future deep-sea research.

The Banda Sea was identified as a suitable core study area of the pilot project to implement I-DREAM programs because it is unique deep-sea (see section 3.1 – 3.4). The next strategic plan is upscaling the project from the Banda Sea (core study area) into larger scales (national or global scale). Upscaling is required to address global change and other environmental threats of deep-sea ecosystem. Some aspects are required for upscaling: 1) inter-institutional and regulatory framework to strengthen institutional networking and collaboration; 2) human resources development especially improving quantity and quality of deep-sea scientists; 3) establishing standardized methodology of data collection and sample analysis are required for data compatibility and data transfer; and 4) national database centre development including metadata and data exchange is needed to optimize the output of deep-sea research (figure 4).

The grand design and its strategic plan of deep-sea research are comprehensive (including all aspects of deep-sea using both literature review and questionnaire survey) and reliable. I-DREAM is a competitive and realistic program to support deep-sea conservation and sustainable use of marine resources to achieve blue economy goals. The limitation of the program and its strategic plan is on their implementations, which require strong long-term political commitment and sustained investment. However, these risks could be minimized if the government has political commitment and invest significant budget in deep-sea research.

4. Conclusions

It is clear that Indonesian deep-sea ecosystem is still underexplored. Data and publication on this ecosystem are very limited and the potential of its biological and natural resources is not well documented. Ironically, this unique ecosystem is threatened by climate change and human activities such as environmental degradation, pollution, and resources extraction. Deep-sea research development faces some obstacles and challenges such as research infrastructure/facilities are very limited and poor institutional management system. Attention and investment in deep-sea research are strongly required to support biological and natural resources security (e.g. food and energy security) and conservation. In order to encourage deep-sea research development, we develop a grand design and innovative strategic plan for deep-sea research and monitoring program. This grand design and strategic plan are comprehensive, competitive, and realistic to be applied in both small and large scales. The Banda Sea is a suitable area for a pilot project location (core area of study) of deep-sea research and then upscaled to a larger area. The grand design and its strategic plan will assist researchers (1) to generate advanced scientific data and information which important to support deep-sea management and decision-policy-making for deep-sea ecosystem, (2) to equip researchers with standardized methods and reliable results of deep-sea research, and (3) meet the growing demand for data and information of deep-sea ecosystem conditions and its potential to achieve blue economic goals including to support conservation actions and human well-being.

Acknowledgments
We thank the two anonymous reviewers for the important feedbacks. We are grateful to Dr Luitgard Schwendennman for valuable input. We thank all respondents for their time to fill the questionnaire and interview.

References

[1] Tomascik T, Mah A J, Nontji A and Moosa M K 1997 *The ecology of the Indonesian seas* part II (Singapore: Periplus Editions) p 746
[2] Rahayu D L 2008 Crustace (*Marine Biodiversity Review of the Arafura and Timor Seas*) eds T Wagey, Z Arifin (Jakarta: ATSEA) p 136
[3] Gordon A L and Susanto R D 2001 Banda Sea surface-layer divergence *Ocean Dynamics* **52**(1) 2-10
[4] Brown D, Ryan P D, Afonso J C, Boutelier D, Burg J P, Byrne T, Calvert A, Cook F, DeBari S, Dewey J F, Gerya T V, Harris R, Herrington R, Konstantinovskaya E, Reston T and Zagorevski A 2011 Arc-continent collision: The making of an orogen *Arc-Continent Collision* vol 4 (*Frontiers in Earth Sciences* vol 4) eds D Brown, P D Ryan (Heidelberg: Springer) pp 477–93
[5] Harris R 2011 The nature of the Banda Arc–continent collision in the Timor region *Arc-Continent Collision* vol 4 (*Frontiers in Earth Sciences* vol 4) eds D Brown, P D Ryan (Heidelberg: Springer) pp 163-211
[6] Rusila-Noor Y, Khazali M and Suryadiputra 2006 *Panduan Pengenalan Mangrove di Indonesia* 2nd rev ed (Bogor: PHKA/WI-IP) p 219
[7] Koropitan A F 2015 Poros Maritim di Era Perubahan Iklim Kompas 25 April 2015 p 7
[8] Minister of Marine Affairs and Fisheries 2009 *Ministerial Decree on Establishing Marine Protected Area in Maluku Province No 69/2009* (Jakarta: KKP) p 8
[9] Badan Pusat Statistik 2013 *Proyeksi Penduduk Indonesia 2010-2035* (Jakarta: Badan Pusat Statistik) p 468
[10] Badan Ketahanan Pangan 2013 *Petunjuk Pelaksanaan Sistem Kewaspadaan Pangan dan Gizi* (Jakarta: Kementerian Pertanian) p 22
[11] Salie M and Handewi P 2002 *Analisis kerawanan pangan wilayah dalam perspektif desentralisasi pembangunan* (Bogor: Pusat Penelitian dan Pengembangan Sosial Ekonomi Pertanian) p 438
[12] Purwantini T B 2014 Pendekatan Rawan Pangan dan Gizi: Besaran, Karakteristik, Dan Penyebabnya *Forum penelitian Agro Ekonomi* **32**(1) 1-17
[13] Rex M A and Etter R J 2010 *Deep-sea Biodiversity: Pattern and Scale* (Harvard University Press) p 354
[14] St John M A, Borja A, Chust G, Heath M, Grigorov I, Mariani P, Martin A P and Santos R S 2016 A dark hole in our understanding of marine ecosystems and their services: Perspectives from the mesopelagic community *Frontiers Mar. Sci.* **3**(31) 1-6
[15] Pezzuto P R, Perez J A A and Wahrlich R 2006 Deep-sea shrimps (Decapoda: Aristeidae): new targets of the deep-water trawling fishery in Brazil *Braz. J. Oceanogr.* **54** 123-34
[16] World Bank and United Nations Department of Economic and Social Affairs 2017 *The Potential of the Blue Economy: Increasing Long-term Benefits of the Sustainable Use of Marine Resources for Small Island Developing States and Coastal Least Developed Countries* (Washington DC: World Bank) p 50

[17] Rengstorf A M, Yesson C, Brown C and Grehan A J 2013 High-resolution habitat suitability modelling can improve conservation of vulnerable marine ecosystems in the deep sea *J. Biogeogr.* **40**(9) 1702-714
[18] Danovaro R, Corinaldesi C, D'Onghia G, Galil B, Gambi C, Gooday A J, Lampadariou N, Luna G M, Morigi C, Olu K, Polymenakou P, Ramirez-Llodra E, Sabbatini A, Sardà F, Sibuet M and Tselepides A 2010 Deep-sea biodiversity in the Mediterranean Sea: The known, the unknown, and the unknowable *PloS ONE* **5**(8) e11832
[19] Gross M 2015 Deep Sea in deep trouble? *Curr. Biol.* **25** R1019-21
[20] Nontji A 2009 *Penjelajahan dan penelitian laut Nusantara dari masa ke masa* (Jakarta: Research Center for Oceanography-LIPI) p 433
[21] Vaz E 2014 Managing urban coastal areas through landscape metrics: An assessment of Mumbai's mangrove system *Ocean Coast. Manag.* **98** 27-37
[22] Silver E, Gill J, Schwartz D, Prasetyo H and Duncan R 1985 Evidence for a submerged and displaced continental borderland, North Banda Sea, Indonesia *Geology* **13**(10) 687-91
[23] Bowin C, Purdy G, Johnston C, Shor G, Lawver L, Hartono H M S and Jezek P 1980 Arc-continent collision in Banda Sea region *AAPG Bull* **64**(6) 868-915
[24] Mazarovich A 2011 Marginal seas—Terminological crisis *Geotectonics* **45**(4) 317-33
[25] Watling L, Guinotte J, Clark M R and Smith C R 2013 *Prog Oceanogr* **111** 91-112
[26] Sukresno B and Kasa I W 2006 Dynamical analysis of Banda Sea concerning with El Nino, Indonesian through Flow and Monsoon by using satellite data and numerical model *Ecotrophic Journal of Environmental Science* **3**(2) 87-91
[27] Polzin K L, Toole J M, Ledwell J R and Schmitt R W 1997 Spatial Variability of Turbulent Mixing in the Abyssal Ocean *Science* **276**(5309) 93-6
[28] Johnson G C 1998 Deep water properties, velocities, and dynamics over ocean trenches *J. Mar. Res.* **56**(2) 329-47
[29] Susanto R D, Gordon A L and Zheng Q 2001 Upwelling along the coasts of Java and Sumatra and its relation to ENSO *Geophys. Res. Lett.* **28**(8) 1599-602
[30] Chamberlin W S and Dickey T D 2007 *Exploring the World Ocean* (New York: McGraw Hill/Higher Education) 394pp
[31] O'Driscoll K T and Kamenkovich V M 2009 Dynamics of the Indonesian seas circulation Part I–The influence of bottom topography on temperature and salinity distributions *J. Mar. Res.* **67**(2) 119-57
[32] Lape P V 2000 Political dynamics and religious change in the late pre-colonial Banda Islands, Eastern Indonesia *World Archaeology* **32**(1) 138-55
[33] Zandvliet K 1998 *Mapping for money: maps, plans and topographic paintings and their role in Dutch overseas expansion during the 16th and 17th centuries* (Amsterdam: Batavian Lion International) p 328
[34] Anwar A 2013 Co-management of marine coastal and local wisdom based kewang as the local government law politics Maluku *Prosiding Seminar Nasional Perkembangan Hukum Lingkungan Kini Dan Masa Depan* eds A Amiruddin, D Imami, I Nurlina, M Priyatna, N Astriani and Y Adharani (Bandung: Padjajaran University) 7pp
[35] Kainama M 2005 Revitalisasi Lembaga Kewang Dalam Rangka Penegakan Hukum Sasi di Maluku Tengah *Jurnal Sasi* **11**(2) 186
[36] Silver J J, Gray N J, Campbell L M, Fairbanks L W and Gruby R L Blue economy and competing discourses in international oceans governance *J. Environ. Dev.* **24**(2) 135-60
[37] Kathijotes N 2013 Keynote: Blue economy-environmental and behavioural aspects towards sustainable coastal development *Proc.-Soc. Behavior. Sci.* **101** 7-13
[38] Sunoko R and Huang H-W 2014 Indonesia tuna fisheries development and future strategy *Mar. Policy* **43** 174-83

[39] Suman A, Irianto H E, Satria F and Amri K 2016 Potensi dan Tingkat Pemanfaatan Sumberdaya Ikan di Wilayah Pengelolaan Perikanan Negara Republik Indonsia (WPP NRI) Tahun 2015 serta Opsi Pengelolaanya *Jurnal Kebijakan Perikanan Indonesia* **8**(2) 97-100

[40] Satria A and Adhuri D S 2010 Pre-existing fisheries management systems in Indonesia, focusing on Lombok and Maluku *Managing Coastal and Inland Waters* eds Ruddle K and Satria A (Dordrecht:Springer) pp 31-55

[41] Gurning S 2006 Analisa konsep Trans-Maluku sebagai sebagai pola jaringan transportasi laut di propinsi Maluku *Proc. Seminar Teori dan Aplikasi Teknologi Kelautan (SENTA) VI* (Surabaya-ITS) 8pp

[42] Lasabda R 2013 Regional Development in Coastal and Ocean in Archipelago Perspective of the Republic of Indonesia: Theoretical Perspective *Jurnal Ilmiah Platax* **1** 92-101

[43] Usman E and Widjaja P H 2012 Morphotectonic analysis at Tanimbar Trench as a base for gas pipe laying between Masela block and Selaru Island, Moluccas Province *Bull. Mar. Geol.* **27**(1) 27-34

[44] Suyadi 2012 Satu dekade kondisi hutan mangrove di Teluk Ambon, Maluku *Jurnal Biologi Indonesia* **8**(1) 197-203

[45] Sprintall J, Gordon A L, Koch-Larrouy A, Lee T, Potemra J T, Pujiana K and Wijffels S E 2014 Indonesian seas and their role in the coupled ocean-climate system *Nat. Geosci.* **7**(7) 487-92

[46] Kementerian Negara Lingkungan Hidup 2014 *Rencana Aksi Nasional Dalam Menghadapi Perubahan Iklim* (RAN-API) eds Rustiadi E (Jakarta: KNLH) p 78

[47] Merki A and Martin S 2015 *Stemming the Tide: Land-based strategies for a plastic-free ocean* (Ocean Conservancy and McKinsey Center for Business and Environment) p 48

[48] Susanto R D, Wei Z, Adi T R, Zheng Q, Fang G, Bin F, Supangat A, Agustiadi T, Trenggono M and Setiawan A 2016 and Martin S 2015 Oceanography Surrounding Krakatau Volcano in the Sunda Strait, Indonesia *Oceanography* **29**(2) 264-72

[49] Aldrian E and Susanto R D 2003 Identification of three dominant rainfall regions within Indonesia and their relationship to sea surface temperature *Int. J. Climatology* **23**(12) 1435-52

[50] Kerns B W and Chen S S 2016 Large-scale precipitation tracking and the MJO over the Maritime Continent and Indo-Pacific warmpool *J. Geophys. Res. Atmos.* **(121)** 8755–776

[51] Susanto R D, Moore T S and Marra J 2006 An ocean color variability in the Indonesian Seas during the SeaWifs Era, *Geochem. Geophys. Geosys.* **7** 1-6

[52] Kinkade C, Marra J, Langdon C, Knudson C, and Ilahude A G 1997 Monsoonal differences in phytoplankton biomass and production in the Indonesian Seas: Tracing vertical mixing using temperature,*Deep Sea Res. Part I* **44** 581–92

IOP Conf. Series: Earth and Environmental Science **184** (2018) 012002 doi:10.1088/1755-1315/184/1/012002

A review of studies on bacteria in the Banda Sea over the past seven decades (1950-2017)

Y Tapilatu[1]

[1]Center for Deep-Sea Research, Indonesian Institute of Sciences (PPLD LIPI), Jl. Y. Syaranamual Guru-guru Poka Ambon 97233, Indonesia

yosmina.tapilatu@lipi.go.id

Abstract. Studies on marine bacteriology in the Banda Sea were started from the 1950s. This paper is an attempt to review all works carried out in the last seven decades. Literature reviews indicated that there was not much research done on marine bacteria in the Banda Sea and adjacent waters such as Ambon Bay. Studies on bacterial biomass, indicator bacteria on water quality monitoring as well as bacteria potential as a producer of secondary metabolites were carried out. However, considering Banda Sea ecosystem that is mostly consisted of deep environments, it would be interesting to prioritize research activities on piezophiles or at least piezotolerant marine bacteria. There is thus an urgent need to determine the priorities in studying marine bacteria in this ecosystem, which should be focused on deep-sea ones.

Keywords: Banda Sea, marine bacteria, deep sea

1. Introduction

The Banda Sea (figure 1) is included in the Coral Reef Triangle Initiative (CTI) area. The CTI area is known to be the global centre of marine biodiversity [1]. The Government of the Republic of Indonesia has issued a Ministerial Decree No. 69/2009 stating the Banda Sea as one of the National Marine Conservation Area [2]. This is because the Banda Sea has natural resources that are yet to be utilized optimally. One of these under-explored resources from the Banda Sea is its marine bacteria.

Marine bacteria refer to the bacterial group that requires 35 ppm NaCl (or the average salinity of seawater) to grow and multiply optimally. Their size ranges from 0.5-5.0 micrometres in length. Spherical bacteria sizes are about 1 micrometre in diameter. This type of microorganisms can be found from coastal waters to the deep ocean. Prior reports [3,4] confirm its existence even at the depth of 10 000 meters of Mariana Trench. They can be found either as free-living, external or internal symbionts to macroorganisms with various level of relationship, from beneficial to pathogenic.

This bacterial group constitutes one portion of the most abundant microorganisms in the sea and has multiple functions in marine ecosystems. They influence the biogeochemical process and play important roles as food for larger predators, as well as the decomposer in the marine trophic web. They also mineralize toxic compounds which enter the sea naturally and/or through human activities. some of them produce natural compounds which are beneficial to humans, particularly in biotechnological applications. Most importantly, marine bacteria are responsible in providing fifty percent of the oxygen that we breathe.

There are many reviews that have been published about marine bacteria in Indonesia [5] and more particularly about those from eastern Indonesian waters [6]. These reviews highlighted on studies

Content from this work may be used under the terms of the Creative Commons Attribution 3.0 licence. Any further distribution of this work must maintain attribution to the author(s) and the title of the work, journal citation and DOI.
Published under licence by IOP Publishing Ltd

about the biotechnology application potential of marine bacteria secondary metabolite from the area. However, to our knowledge none has been done especially about those carried out in the Banda Sea and adjacent waters under its influence, i.e Ambon Bay. This paper is an attempt to provide a comprehensive literature review, with the objective to provide the most updated review on marine bacteria studies carried out in the Banda Sea and Ambon Bay. It was expected to be a reference for future studies that will be carried out regarding marine bacteria in the aforementioned area.

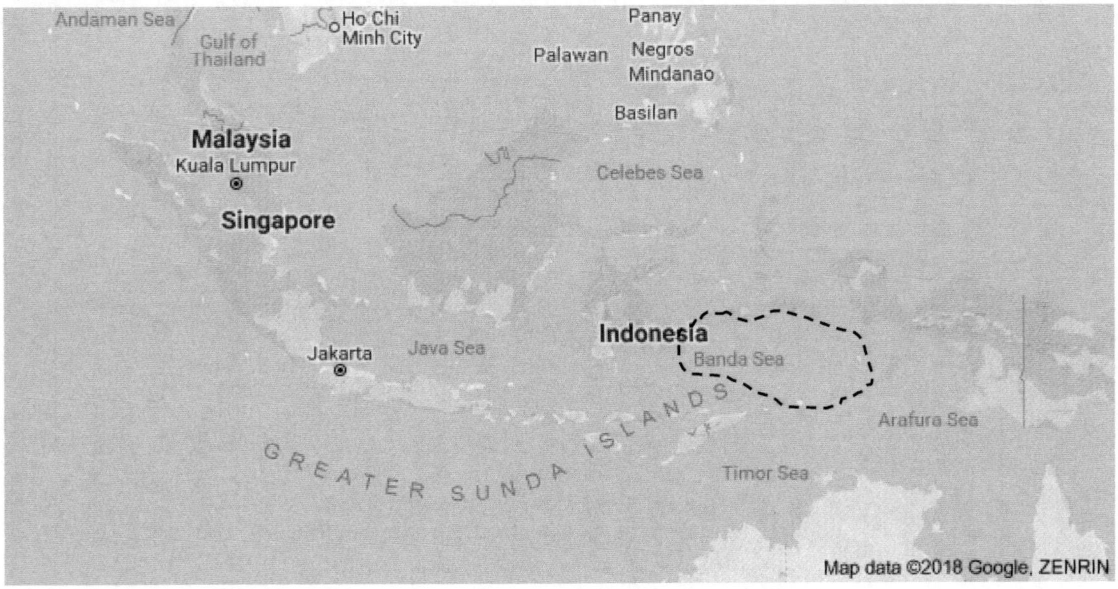

Figure 1. Maps of Indonesia. The dashed line indicates the geographic area of the Banda Sea.

2. Methodology

Previously published literature contained information on marine bacteria studies in the Banda Sea from the last seven decades (1950-2016) were used as primary sources. Google search engine was used to initially track published literature available electronically in English and/or Indonesian languages, using combined keywords such as "marine bacteria banda sea" and "marine bacteria ambon bay". This was because most of the published works were either written in Indonesian, or few of them were indexed by reputable scientific literature search engine. The results obtained served as starting point for further literature search, mainly by searching for reference lists in reviews and original research articles. Articles which were not available electronically were hand-searched in relevant journals available in our institution's library. There were several minimum criteria for inclusion. The first criterion included literature published in international, national and local journals/proceedings and graduate thesis. The full-text availability of literature obtained, and the relevance of the results presented with the context of this paper constituted the second and the third criteria.

3. Results and discussion

3.1. Research history and challenges

There were only eight articles published about original research on marine bacteria from the Banda Sea and Ambon Bay during the last seven decades (1950-2017) (Table 1), which means that there was -in average- one article published every decade. This was very low in quantity compared to studies on marine bacteria in other parts of Indonesia. For example, literature search on line back in 2011 [5] and recently indicated that there were at least 27 original research articles published about marine bacteria studies in other parts of Indonesia in the period of 1970 to 2010.

IOP Conf. Series: Earth and Environmental Science **184** (2018) 012002 doi:10.1088/1755-1315/184/1/012002

Table 1. Summary of bacterial studies in the Banda Sea, 1950-2017.

No	Sampling area	Study type	Methods used	Key findings/results	Publication year	Reference
1	Banda Sea	Exploration of deep-sea bacteria	Conventional isolation using selective media	Isolation of deep-sea starch hyrdolyzers, nitrate reducers, sulphate reducers, ammonifiers and nitrifiers bacteria	1959	[7]
2	Ambon Bay	Indicator bacteria (coliform, faecal coli and faecal *Streptococcus*)	Most Probable Number (MPN)	Lower number in Ambon Bay compared to Jakarta Bay and Banten Bay	1988	[8]
3	Banda Sea	Marine bacterial ecology	Adenosine Tri Phosphate (ATP) measurement of two different periods	Bacterial numbers were not affected by the monsoon period	1990	[9]
4	Banda Sea	Marine bacterial ecology	Indirect measurement of cell numbers using Phycoerythrin absorbance assessment	*Synechococcus*-type species prefer deeper water column (twilight zone) and indicated diurnal pattern based on their low surface abundance	1990	[10]
5	Ambon Bay	Indicator bacteria	MPN	Detection of coliform total, fecal coliform and *E. coli* at Lata village waters	1996	[11]
6	Ambon Bay	Hydrocarbon-degrading bacteria (HCB) number assessment	MPN	Higher density of HCB number at Ambon Inner Bay than at Ambon Outer Bay	1997	[12]
7	Ambon Bay	Heterotrophic bacteria	TPC	Higher density of heterotrophic bacteria at Ambon port waters, due to anthropogenic activity, with vertical distribution pattern in accordance with dissolved oxygen concentration in Ambon Bay	1998	[13]
8	Ambon Bay	Exploration of secondary-metabolite producer bacteria	Conventional isolation using selective media and initial assessment of the secondary metabolite produced profiles	Isolation of putative agarolytic bacteria, exopolysaccharide-producing bacteria and marine Actinomycetes	2015	[14]

This accounted for about six articles per decade in forty years. Five from eight published studies (Table 1) were carried out at Ambon Bay area. Three main subjects were treated, namely exploration of bacteria with potentially important functions (either ecologically or commercially) [7,14], marine bacterial ecology [9,10] and indicator bacteria as a part of Ambon Bay water quality monitoring [8,11-13]. These subjects were less varied compared to marine bacterial research carried out in other parts of Indonesian waters, with also varied methodologies (c.f [5] and references therein).

It is interesting to note, however, that the earliest study about marine bacteria in the Banda Sea was also the first marine bacteria study ever published that was carried out in Indonesian marine environment. The study was reported in the late 1950s, as part of the Danish world expedition, Galathea 2 that took place in 1951. Weber Deep, Banda Sea was one of the stations where ZoBell and Morita collected sediment samples from 7 250 m for subsequent incubations and attempt of isolations of putative *in situ* deep-sea bacteria [7]. They were the first to report the presence of starch hydrolyzers, nitrate reducers, sulphate reducers, ammonifiers and nitrifiers that were active in the differential media at 700 atm, and suggested that these were indigenous species which were probably limited by the nutrient content of sea water and sediment and not by high hydrostatic pressure or low temperature. Further literature research indicated that there was no record of continued studies regarding this sample afterward. Nevertheless, this was one of the seminal works on deep-sea bacteria globally afterward.

Research activities on marine bacteria in the Banda Sea have practically halted afterward and restarted in the 1990s. Whereas in Ambon Bay particularly, marine bacteria studies were started by Thayib and Razak in 1988, when they carried out observations on bacterial indicator on Ambon Bay water quality compared to Jakarta Bay and Bay of Banten [8]. The reason for this long gap of activities could possibly due to lack of proper infrastructures to carry out the research in this field at that period. The Snellius II Expedition in 1984-1985 resulted in two original research articles on marine bacteria. The first one by Vosjan et al. [9] investigated on the microbial biomass, respiratory activity and bacterial numbers in the Banda Sea as a function of the two monsoon periods, northwest and southeast. The ATP measurement as relative biomass of less than 50 µm microorganism at the eastern part of Banda Sea indicated that it was not affected by the monsoon period. A similar trend was observed also when the respiratory activity was measured. They also found out that the bacterial numbers tended to increase during the night.

The second study was done by Zevenboom, who carried out research on the relative abundance of red-pigmented picocyanobacteria during August 1984 and February/March 1985 [10]. It was observed in this study that *Synechococcus*-type species showed a preference for deeper layers where light intensities were low, and indicated diurnal pattern based on their low surface abundance. Unfortunately, no further studies were carried out to follow up the results obtained. Yet these were few of the rare studies on bacterial ecology in the Banda Sea.

As adjacent marine environment that is heavily influenced by the Banda Sea, Ambon Bay became one of the locations where research activities were carried out intensively in the 1990s. This included the few studies on water quality indicator [11], hydrocarbon-degrading [12] and heterotrophic [13] bacteria, carried out from 1996 to 1998 in Ambon Bay coastal waters. Unfortunately, all research activities were suspended due to civil unrest from 1999 to 2004.

Studies on marine bacteria in Ambon Bay were restarted in around 2010, on water quality indicators and continued until 2016. In parallel, preliminary studies about marine bacteria from this area and their potential as producers of bioactive compounds were conducted (cf. [14]).

Nevertheless, no studies were carried out on marine bacteria in the Banda Sea after 2004 to our knowledge. This was due to several factors, i.e. lack of adequate funding, infrastructure, and competent human resources. Sampling effort requires substantial fund, as expedition in the Banda Sea is costly, in terms of logistic and technical details to accommodate all researchers involved. With its vast area, sampling in the Banda Sea cannot be accomplished in a single expedition of ten cruising days, as were carried out in the last five years by Center for Deep-Sea Research, Indonesian Institute of Sciences (LIPI). As they can only be seen using microscopes and takes hours of bench works and expensive culture media ingredients, marine bacteria are not one of the priorities for the research to be funded, in contrast with other marine organisms that are easily visible, colourful and highly publicized.

Other than that, lack of proper and basic infrastructures in marine bacteria hinder studies in this field on Banda Sea ecosystem. For instance, as per author's experience, it is difficult to obtain adequate storage facilities for samples, let alone isolates, such as -80°C freezer or liquid nitrogen container in Maluku Province's capital, Ambon. This is not only due to the limited funding in government institutes research equipments purchase but also in the absence of seller or dealer of substantial provision such as liquid nitrogen.

The limited funding and infrastructures combined with limited competent human resources in the field worsen the situation. Informal discussions carried out with the local undergraduate students indicated that they feel discouraged to pursue interest in marine bacteria studies because of the long hours spent in laboratory works and the expensive culture media or molecular biology kits and disposable items.

3.2. Strategies for future research

Applied bacterial studies are important aspects that should be put forward in the future orientation of research in the Banda Sea. As previously mentioned in the introduction, the development of marine and fisheries sector in Banda Sea ecosystem is important for Indonesia, in particular for the people of Maluku Province. Limited infrastructures, research funds and competent human resources require the following strategies: 1) Prioritization of the understudied deep-sea bacteria research, notably the exploration of those with potential application in biotechnology; 2) Provision of the existing laboratories with adequate equipment for sampling, analysis, and preservation of deep-sea bacteria; 3) Provision of multi-years funding for the community structure analysis and bioprospecting works; 4) Preparation of the critical mass of researchers in the field of marine bacteria.

The first strategy is needed to optimize the utilization of the limited available resources. Marine microbes in the Banda Sea have yet to receive the attention they deserve as one of, if not the most, important actors in this environment. However, considering the challenges faced as detailed in the previous section (3.1), prioritization is necessary to ensure that the available resources are put to good use. Other than what has been done by Schwartz and Morita in the 1950-s, not much if any is known about the role of deep-sea bacteria in general in Banda Sea ecosystem. Specific studies such as bacterial community structure need to be carried out to gain better understanding of these roles. In parallel, considering the great potential of deep-sea bacteria of Banda Sea as producers of beneficial compounds, bioprospecting works could be also carried out. An unpublished report of preliminary attempts in isolating putative exopolysaccharide-producing bacteria from Ambon Bay sediment (400 m) [15] yielded one isolate capable of producing such compound. Bacterial exopolysaccharide has wide range application in biotechnology, including pharmacy [16].

Adequate equipments for sampling, analysis, and preservation of deep-sea bacteria in relation with the first strategy will thus be indispensable. This could be done in multiple years and could be started with basic equipments such as -80°C freezer, liquid nitrogen container, pressurized microbial incubation chamber or adaptable shakers. The first two are needed for isolate storage. The pressurized microbial incubation chamber will be useful to mimic the living condition in the deep sea. The adaptable shaker will be necessary to ensure the provision of oxygen inside the pressurized chamber.

The third strategy (continual funding) is essential in ensuring that the work carried out would not be done only in initial phases and had to be stopped due to the lack of funding, but would be accomplished thoroughly. This could be secured not only from internal funding but also through collaborations with other institutes or universities, in national and international levels.

In regard with the fourth strategy, human resource with skills in marine bacteria is lacking in Maluku Province. As a consequence, efforts should be done to raise young and/or potential researchers' awareness on the importance of advancing studies on bacteria from Banda deeper waters.

There is a need of representing the marine bacteria subject as not an intimidating one, and this would take further socialization and communication works to the young generation early on.

4. Conclusion

There is still a lot of work to be done about marine bacteria in the Banda Sea. Bacteria from Banda deeper waters are still under explored. Challenges in marine bacteria study in the Banda Sea require strategies that should be carried out to ensure accomplishment of necessary research works. These studies would not only contribute to the basic knowledge about their presence and potential roles in Banda Sea ecosystem but also could lead to the discoveries of potential producers of beneficial compounds. These future studies could only be possible within regional and international collaborative research scheme, as large numbers of infrastructures, as well as equipment and human resources, will be needed to accomplish them.

References

[1] Burke L, Reytar K, Spalding M and Perry A 2011 *Reefs at risk. Revisited* (Washington DC: World Resources Institute) p 124

[2] Indonesian Minister of Marine Affairs and Fisheries 2009 *Ministerial Decree on the Marine Protected Area Establishment in Maluku Province No 69/2009* [Internet] accessed on Apr 12th, 2017 from http://kkji.kp3k.kkp.go.id/index.php/dokumen/regulasi-hukum/keputusan-menteri/finish/14-ke-putusan-menteri/187-kepmen-kp-no-kep-69-men-2009-tentang-pene-tapan-kawasan-konservasi-perairan-nasional-laut-banda-di-provinsi-maluku

[3] Kato C, Li L, Nogi Y, Nakamura Y, Tamaoka J and Horikoshi K 1998 Extremely barophilic bacteria isolated from the Mariana Trench, Challenger Deep, at a depth of 11,000 meters *Appl. Environ. Microbiol.* **64** 1510–13

[4] Deming J W, Somers L K, Straube W L, Swartz D G and MacDonald M T 1988 Isolation of an obligately barophilic bacterium and description of a new genus, *Colwellia* gen. nov *Syst. Appl. Microbiol.* **10** 152–60

[5] Tapilatu Y 2011 Marine prokaryotes in Indonesia: research history and potential use in bio-technology *Proc. 1st Int. Symp. for Sustainable Humanosphere* Oct 3, 2011 (Ambon: R & D Unit for Biomaterials-LIPI) pp 21–9 doi: 10.13140/2.1.3345.8563

[6] Tapilatu Y 2016 Marine bacteria from eastern Indonesia waters and their potential use in biotechnology *Omni Akuatika* **12**(1) 81-5 doi: http://dx.doi.org/10.20884/1.oa.2016.12.1. 32

[7] Zobell C E and Morita R Y 1959 Deep-Sea Bacteria *Galathea Report* **1** 139–54

[8] Thayib S S and Razak H 1988 Observation on the bacteriological indicator, heavy metal and pesticide at Ambon Bay, Banten Bay and Jakarta Bay, *Perairan Indonesia: Biologi, Budidaya, Kualitas Perairan dan Oseanografi*, pp 124–31

[9] Vosjan J H, Nieuwland G and Ruyitno 1990 Monsoonal differences in microbial biomass, respiratory activity and bacterial numbers in the Banda Sea *Neth. J. Sea Res.* **25**(4) 501–11

[10] Zevenboom W 1990 Picocyanobacteria in the Banda Sea during two different monsoons *Neth. J. Sea Res.* **25**(4) 513–21

[11] Indah L and Chasanah E 1996 Kualitas Mikrobiologi Perairan Pantai Teluk Ambon Bagian Dalam Ditinjau Dari Segi Kesehatan Lingkungan *Prosiding Seminar dan Lokakarya Pengelolaan Teluk Ambon* (Ambon: LON LIPI) pp 56–62

[12] Sutiknowati L I 1997 Sebaran Kepadatan Bakteri Pemecah Minyak di Teluk Ambon *Seminar Kelautan LIPI–UNHAS* (Makassar: LON LIPI–UNHAS) pp 260–64

[13] Sutiknowati L I 1998 Abundance and distribution of heterotrophic bacteria in Ambon Bay (a preliminary study) *Perairan Maluku dan Sekitarnya* **12** 17–24

[14] Tapilatu Y 2015 Isolation of marine bacteria in Ambon Bay with potential biotechnological features *Taprobanica* **7**(3) 143-50

[15] Tapilatu Y 2012 In search of exopolysaccharide–producing bacteria from Maluku deep–sea area *Technical Report L'Oreal–UNESCO Indonesian National Fellowship for Women in Science 2011* p 4

[16] Tapilatu Y 2015 Status of Drug Discovery Research Based on Marine Organisms from Eastern Indonesia *Proc. Chem.* **14** 484–92

The Banda Sea: a hotspot of deep-sea echinoderms diversity in Indonesia

A Setyastuti[1] and I Wirawati[1]

[1]Research Center for Oceanography-Indonesian Institute of Sciences (LIPI), Jl. Pasir Putih 1, Ancol Timur, Jakarta 14430 Indonesia

ana.setyastuti@gmail.com

Abstract. A list of Echinodermata fauna recorded from the Banda Sea has been compiled from the literature published between 1889-1995 as well as from the LIPI-RCO reference collection. To date, 44 species from five extant classes (Asteroidea, Ophiuroidea, Crinoidea, Echinoidea, and Holothuroidea) have been recorded from this area that consisting of the epipelagic (< 200 m depth), mesopelagic (200-1000 m depths) and bathypelagic (> 1000 m depth) species. Research history, taxonomy, and depth distribution are briefly discussed here. However, deep-sea habitats all over the world remain unexplored, including the Banda trench. The opportunity to discover both new records and new species for science from this trench is still wide open since the hitherto record of its marine biodiversity is still a few compared to the discovery on the shallow water. More deep-sea exploration will be needed to improve our knowledge of marine biodiversity.

Keywords: Asteroidea, Ophiuroidea, Crinoidea, Echinoidea, Holothuroidea, Banda trench, research history.

1. Introduction

The Echinodermata group is characterized not only by the spiny integument as its phylum definition, but also by a combination of adult pentaradial symmetric body, a calcite skeleton, and water vascular system. The phylum is an ancient group of exclusively marine invertebrate that was defined for the first time by Herouard in 1899 [1] which includes five extant class i.e. crinoids (feather stars and sea lilies), asteroids (sea stars), echinoids (sea urchins, heart urchins and sand dollars), ophiuroids (brittle stars and basket stars) and holothuroids (sea cucumbers). Echinoderms species are mostly free-living and benthic; the distributions are from tropical to the sub-tropical area with a depth range from shallow water to deep sea.

The Echinodermata is well represented in Indonesian waters. However, the comprehensive lists of its species in Indonesia have never been published. Most of the publications about Indonesian Echinodermata biodiversity were reported partially based on location (figure 1). For example diversity in Anambas waters [2], Seribu Island waters [3], East Java waters [4], Nusa Tenggara waters [5-7], Spermonde waters [8], Lembeh Strait waters [9] and Maluku waters [10-20].

Content from this work may be used under the terms of the Creative Commons Attribution 3.0 licence. Any further distribution of this work must maintain attribution to the author(s) and the title of the work, journal citation and DOI.
Published under licence by IOP Publishing Ltd

IOP Conf. Series: Earth and Environmental Science **184** (2018) 012003 doi:10.1088/1755-1315/184/1/012003

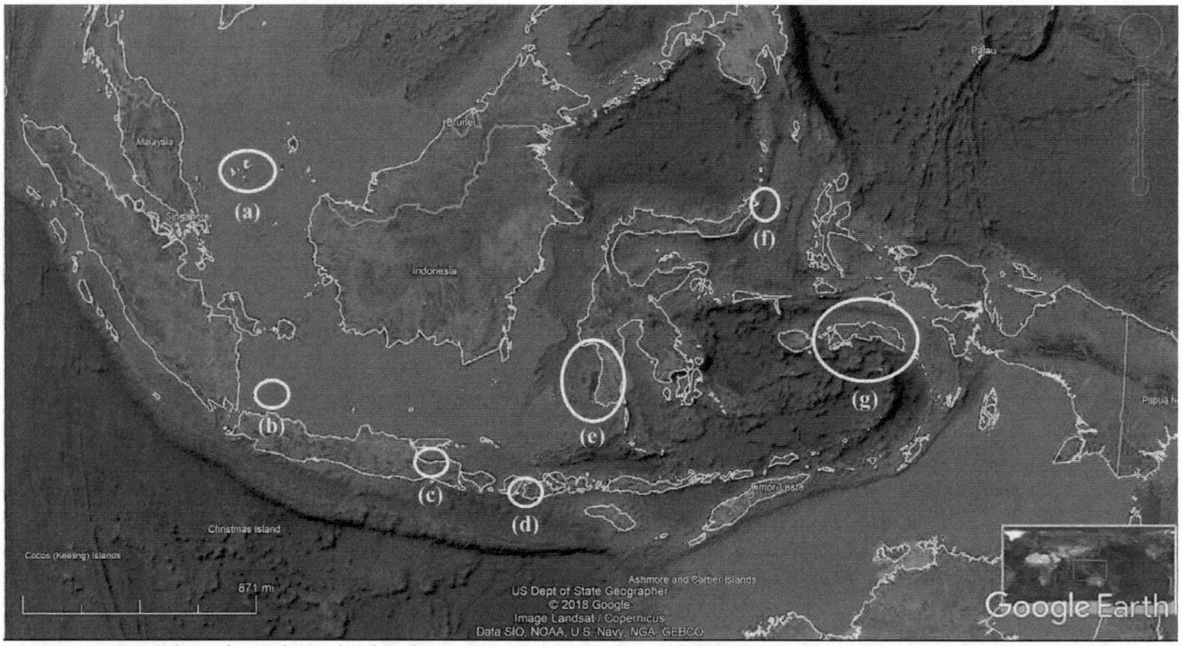

Figure 1. Locations in Indonesia that have taxonomical data of the Echinodermata: a: Anambas Island; b: Seribu Island, West Java; c: East Java; d: Nusa Tenggara; e: Spermonde Island, South Sulawesi; f: Lembeh Strait, North Sulawesi; g: Maluku.

The Indonesian marine life diversity is hard to be precisely reported not only due to the vast area to discover but also to the variety of habitats to be explored. Indonesia comprises of more than 18,000 islands and coastline of about 109,000 km [23]. Indonesia waters composed of many ecosystem varieties from the shallow water to the deep sea i.e. mangrove, seagrass, coral reef, and deep water. The Banda Sea is one of the important deep-sea realm in Indonesia which is not well documented from any viewpoints including its species diversity and energy resources [23].

The study on deep-sea echinoderm in Indonesia has received minor attention for centuries. The most recent overview of the Zoological Catalogue of Australia [1] enlisted several type-species from the Banda Sea that were deposited in the Australian Museum. We used the bibliography noted in this catalog as a reference in tracing the exploration history of the Banda Sea.

2. Data collection

The substance of the discussion in this manuscript was based on the data from any publication which had a topic "Echinodermata and the Banda Sea." Ten publications were successfully traced. They were one overview paper [1] and old taxonomic papers for the rest [24-31]. To comprehend the data, we used also the record of echinoderms specimens from the Banda Sea that were deposited in the LIPI-Recearch Center for Oceanography (RCO) Reference Collection (Refcoll) [32].

3. Results and Discussion

A total of 44 species was collected from the Banda Sea that belongs to five classes. 35 species were successfully listed based on the literature and the rest of nine species were the record from the LIPI-RCO Refcoll database (Table 1). It was comprised of 10 species of Holothuroidea, 12 species of Asteroidea, and 14 species of Ophiuroidea. Echinoidea and Crinoidea possessed four species each respectively (figure 2).

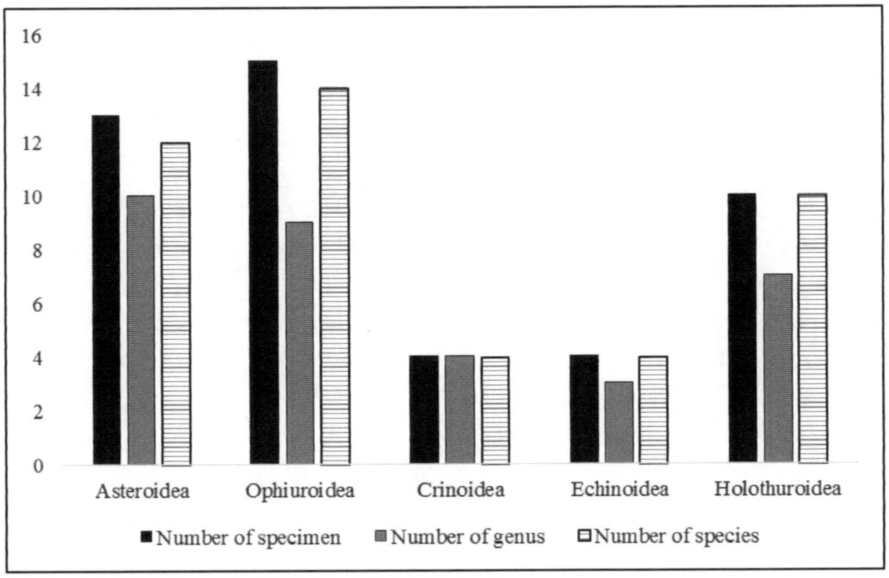

Figure 2. The number of the specimens, genus, and species of Echinodermata class found in the Banda Sea to date.

3.1. Research history

It can be seen from the references [1,24-31] that all species observed from the deepest levels of the Banda Sea were collected from the date far back after G.E. Rumphius exploration [33]. Meanwhile, records from specimens deposited in the LIPI-RCO Refcoll [32] shows that it was from the exploration year 1975, Rumphius Expedition II funded by UNESCO [34]. Sladen [24] through Challenger Expedition describe almost all Asteroidea specimen from the Banda Sea as new species. All species described by him were taken from the seafloor of the Banda Sea using dredging equipment; hence, the depth recorded in the list is remaining constant at a level deeper than 200 meters. Doderlain [28-31] also described the species of Asteroidea, of which some species are the same species found by Sladen [24]. Specimens identified by Doderlain were collected using dredge and corer during the Siboga expedition. He specifically described the starfish specimens. His publication in 1917 was focused on the genus of *Astropecten* [28] and three years later he described the genus of *Luidia* [29]. In 1924, he noted the asteroid species of Pentagonasteridae family [30] and he continued the description of asteroid subfamily of Oreasterinae in 1936 [31]. Holothuroidea and Echinoidea species listed here mostly from the reference of Sluiter [25] and De Meijere [26] which also resulted from Siboga expedition. Meanwhile, Clark [27] re-record one of Holothuroidea species that previously described by Sluiter [25]. Overall, the large number of new species from the deep water of the Banda Trench described by the aforementioned authors in the past is not surprising because the Banda Sea at that time had been wholly unexplored. Moreover, the sampling method using dredge or corer had also never previously been employed.

IOP Conf. Series: Earth and Environmental Science **184** (2018) 012003 doi:10.1088/1755-1315/184/1/012003

Table 1. Species list of Banda Sea echinoderms, based on literatures review. e: epipelagic (0-200 m), m: mesopelagic (200-1000 m), b: bathypelagic (> 1000 m)

No	Class/Family	Species present status[a]	Cited as	Depth	Synonim	Ref.
Asteroidea						
1	Benthopectinidae	*Cheiraster (Luidiaster) teres* Sladen, 1889	*Cheiraster teres* (Sladen, 1889)	m	*Pontaster teres* Sladen, 1889	[1,24]
2	Astropectinidae	*Astropecten acanthifer* Sladen, 1883	*Astropecten acanthifer* Sladen, 1883	m	*Astropecten acanthifer* Sladen, 1883	[24,28]
3	Astropectinidae	*Astropecten bandanus* Döderlein, 1917	*Astropecten bandanus* Döderlein, 1917	e, m	*Astropecten bandanus* Döderlein, 1917	[1,28]
4	Astropectinidae	*Astropecten polyacanthus* Müller & Troschel, 1842	*Astropecten polyacanthus* Müller & Troschel, 1842	e	*Astropecten hystrix* Müller & Troschel, 1842	[28]
5	Pseudarchasteridae	*Paragonaster ctenipes* Sladen, 1889	*Paragonaster ctenipes* Sladen, 1889	e, m	*Paragonaster ctenipes* Sladen, 1889	[1,24,30]
6	Goniasteridae	*Rosaster symbolicus* [30]	*Nymphaster symbolicus* var. Breviradiata Sladen, 1889	e, m	*Nymphaster symbolicus* Sladen, 1889	[1,24]
7	Ophidiasteridae	*Tamaria tumescens* Koehler, 1910	*Tamaria tumescens* Koehler, 1910	e, m	*Ophidiaster tumescens* Koehler, 1910; *Tamaria ajax* Livingstone, 1932; *Tamaria propetumescens* Livingstone, 1932	[1]
8	Zoroasteridae	*Pholidaster squamatus* Sladen, 1889	*Pholidaster distinctus* Sladen, 1889	m	*Pholidaster squamatus* Sladen, 1889	[24]
9	Luidiidae	*Luidia avicularia* Fisher, 1913	*Luidia avicularia* Fisher, 1913	e	*Luidia avicularia* Fisher, 1913	[29]
10	Oreasteridae	*Protoreaster nodosus* Linnaeus, 1758	*Oreaster turritus* Gray, 1840	e	*Pentaceros turritus* Gray, 1840	[31]
Ophiuroidea						
11	Asteroschematidae	*Ophiocreas sibogae* Koehler, 1904	*Ophiocreas sibogae* Koehler, 1904	e, m, b	Ophiocreas longipes Mortensen, 1924	[1]
12	Ophiohelidae	*Ophiomyces delata* Koehler, 1904	*Ophiomyces delata* Koehler, 1904	m, b	*Ophiomyces delata* Koehler, 1904	[1]
13	Ophiodermatidae	*Bathypectinura heros* (Lyman, 1879)	*Bathypectinura heros* (Lyman, 1879)	m, b	*Pectinura heros* Lyman, 1879; *Pectinura lacertosa* Lyman, 1883; *Pectinura tessellata* Lyman, 1883; *Pectinura conspicua* Koehler, 1897; *Ophiocrates lenta* Koehler, 1904; *Pectinura modesta* Koehler, 1904; *Ophiocrates secundus* Koehler, 1906; *Pectinura elata* Koehler, 1906; *Athypectinura gotoi* Matsumoto, 1915; *Phiozonella brachyactis* Clark, 1939; *Ophiocrates intervallus* Madsen, 1947	[1]
14	Ophiodermatidae	*Ophiopsammus yoldii* (Lütken, 1856)	*Ophiopsammus yoldi* (Lütken, 1856)	e, m	*Ophiopeza yoldii* Lütken, 1856; *Ophiopeza conjungens* Bell, 1884; *Ophiopeza exilis* Koehler, 1905	[1]
15	Ophiuridae	*Ophiopallas paradoxa* Koehler, 1904	*Ophiopallas paradoxa* Koehler, 1904	e, m		[1]
16	Ophiotrichidae	*Ophiogymna elegans* Ljungman, 1866	*Ophiogymna elegans* Ljungman, 1866	e	*Ophiocampsis inermis* Koehler, 1905	[1]

continue on the next page

IOP Conf. Series: Earth and Environmental Science **184** (2018) 012003 doi:10.1088/1755-1315/184/1/012003

Table 1. Continued

No	Class/Family	Species present status[a]	Cited as	Depth	Synonim	Ref.
17	Ophiotrichidae	*Ophiothrix (Acanthophiothrix) armata* Koehler, 1905	*Ophiothrix (Acanthophiothrix) armata* Koehler, 1905	e	*Ophiothrix armata* Koehler, 1905	[1]
18	Ophiuridae	*Amphiophiura insolita* Koehler, 1904	*Amphiophiura insolita* Koehler, 1904	m, b	*Ophiura monaria* Clark, 1949	[1]
Crinoidea						
19	Comatulidae	*Capillaster sentosus* Carpenter, 1888	*Capillaster sentosa* Carpenter, 1888	e	*Actinometra sentosa* Carpenter, 1888	[1]
20	Comatulidae	*Clarkcomanthus littoralis* Carpenter, 1888	*Clarkcomanthus littoralis* Carpenter, 1888	e	*Actinometra littoralis* Carpenter, 1888	[1]
21	Comatulidae	*Phanogenia multibrachiata* Carpenter, 1888	*Phanogenia multibrachiatus* Carpenter, 1888	e	*Actinometra multibrachiata* Carpenter, 1888	[1]
22	Comatulidae	*Stephanometra indica* Smith, 1876	*Stephanometra indica* Smith, 1876	e	*Comatula indica* Smith, 1876; *Antedon spicata* Carpenter, 1881; *Antedon monacantha* Hartlaub, 1890	[1]
Echinoidea						
23	Saleniidae	*Salenocidaris hastigera* A. Agassiz, 1879	*Salenocidaris hastigera* A. Agassiz, 1879	m, b		[1]
24	Cidaridae	*Eucidaris metularia* Lamarck, 1816	*Cidaris metularia* Lamarck, 1816	e		[26]
25	Temnopleuridae	*Temnotrema bothryoides* L. Agassiz in L. Agassiz & Desor, 1846	*Pleurechinus bothryoides* L. Agassiz in L. Agassiz & Desor, 1846	e		[26]
Holothuroidea						
26	Holothuriidae	*Pearsonothuria graeffei* [43]	*Holothuria graeffei* Semper, 1868	e		[25]
27	Holothuriidae	*Holothuria (Theelothuria) kurti* Ludwig, 1891	*Holothuria kurti* Ludwig, 1891	e		[25]
28	Phyllophoridae	*Phyllophorella longipeda* Semper, 1867	*Cucumaria longipeda* Semper, 1867	e		[25]
29	Cucumariidae	*Cucumaria vilis* Sluiter, 1901	*Cucumaria vilis* Sluiter, 1901	e		[25]
30	Sclerodactylidae	*Afrocucumis africana* Semper, 1867	*Pseudocucumis africana* Semper, 1867	e		[25]
31	Synaptidae	*Synaptula reticulata* Semper, 1867	*Chondrocloea reticulata* Semper, 1867	e		[25]
32	Holothuriidae	*Holothuria (Microthele) whitmaei* Bell, 1887	*Mulleria maculata* Brandt, 1835	e		[25]
33	Stichopodidae	*Thelenota ananas* (Jaeger, 1833)	*Stichopus ananas* Quoy & Gaimard, 1833	e		[25]
34	Synaptidae	*Synaptula lactea* (Sluiter, 1887)	*Synaptula lactea* (Sluiter, 1887)	e	*Synapta lactea* Sluiter, 1887	[25, 27]
35	Synaptidae	*Synaptula lamperti* Heding, 1928	*Synaptula lamperti* Heding, 1928	e	*Synaptula membrana* Heding, 1928; *Synaptula purpurea* Heding, 1928	[1]

[a] updated from [35].

3.2. Taxonomy and depth distribution

Several species in the older literature had multiple different synonyms because of uncertain identification and unfixed systematic in the past (Table 2). Systematic problems because of minor classificatory and nomenclatorial changes are unavoidable. More investigations by using advanced identification methods will lead to more invention of species characteristic. Thus will strength the systematics of the taxa. However, at present, by tracing through WORMS [35] (The World Register of Marine Species) as one of the trusted and reputable online sources for marine species systematics, all species enlisted here are now recognized as valid.

The echinoderm species found in the Banda Sea enlisted herein are divided into three major depth (epipelagic, mesopelagic and bathypelagic). Epipelagic depth is species found in the depth less than 200 m. Mesopelagic depth is species found in the depth between 200-1000 m. Bathypelagic depth is species found in the depth more than 1000 m. Almost all of the species described in old taxonomical reference [1], [24-31] (Table 1) were discovered from the zone range of mesopelagic and bathypelagic. On the other hand, all species from the data of RCO Refcoll (Table 2, [32]) were discovered from the epipelagic depth, even shallower. Species of Holothuroidea inhabiting the shallower bottom than other Echinodermata class. Based on a number of Echinodermata species inhabiting the deeper water, Ophiuroidea has the largest species diversity, continue to Asteroidea, Crinoidea, and Echinoidea respectively. It confirms that Echinodermata species has a wide range of depth distribution habitat from shallow water to the deepest sea such as the Banda Trench.

Table 2. Species list of Banda Sea echinoderms, based on specimen collection. Voucher: specimens deposited at LIPI-RCO Refcoll. e: epipelagic (0-200 m), m: mesopelagic (200-1000 m), b: bathypelagic (> 1000 m)

No	Class/Family	Species	Voucher	Depth	Notes	Reference
Asteroidea						
1	Goniasteridae	*Fromia indica* (Perrier, 1869)	ECH 0694, ECH 0961	e		[32]
2	Ophidiasteridae	*Linckia multifora* (Lamarck, 1816)	ECH 0699	e		[32]
Ophiuroidea						
3	Ophiotrichidae	*Ophiothrix* Sp	ECH 0743	e		[32]
4	Ophiotrichidae	*Ophiothrix nereidina* (Lamarck, 1816)	ECH 0768, ECH 1280	e		[32]
5	Ophiotrichidae	*Ophiothrix ciliaris* Müller & Troschel, 1842	ECH 0875	e		[32]
6	Ophiotrichidae	*Ophiothrix rhabdota* H.L. Clark, 1915	ECH 0895	e		[32]
7	Ophiotrichidae	*Ophiothrix deceptor* Koehler, 1922	ECH 1243	e	lay on *Acropora*	[32]
8	Ophiotrichidae	*Macrophiothrix longipeda* (Lamarck, 1816)	ECH 1277	e		[32]
Echinoidea						
9	Echinometridae	-	ECH 1234	e	unfixed identification	[32]

3.3. Foresight of the Banda Trench

The opportunity to explore the Banda Trench as one of the deepest seas of the world is still wide open since the hitherto record of its marine biodiversity remains limited. Our knowledge of echinoderms biodiversity is inadequate, and knowledge of already described diversity also is limited [36]. More biodiversity survey on the deep sea will improve our knowledge of the benthic invertebrate.

Numerous species remain difficult or impossible to get the accurate identification. This leads us to study more on advanced identification methods. Indonesia only has limited taxonomist, particularly who is a concern on megabenthic invertebrate such Echinodermata. The chance of multinational cooperation on exploring the deep water of Banda Trench will remain wide open in the upcoming year.

4. Conclusion

This review has highlighted the gaps in terms of the deep-sea echinoderms research that awaits to be carried in the future. Research using dredge or corer up to the bathypelagic or abyssopelagic (> 4000 m depth) zone is the attractive issue for the next research methods. Multinational collaboration also should become a necessary thing to do in the upcoming years because discovery surveys remain chaotic for many decades. However, the role of taxonomy study will still become the main subject for the next years. Thus our responsibility together to discover as much as possible species diversity before they got vanish.

Acknowledgments

The author thanks the two anonymous reviewers for the invaluable input to the manuscript.

References

[1] Rowe F and Gates J 1995 *Zoological Catalogue of Australia*, in *Echinodermata* (Melbourne: CSIRO Australia) 33 p 578

[2] Purwati P and Lane D 2004 Asteroidea of the anambas expedition 2002. *The Raffles Bulletin of Zoology.* **11** 89-102

[3] Pratiwi F 2011 *Inventarisasi jenis-jenis holothuroidea (echinodermata) di rataan terumbu beberapa pulau taman nasional kepulauan seribu, jakarta* (Bachelor thesis of Mathematic and Science Department: University of Indonesia, Jakarta) p 69

[4] Purwati P and Wirawati I 2012 Sea cucumbers of Teluk Prigi, Southern coast of East Java Province *Oseanologi dan Limnologi di Indonesia* **38 (2)** 241-54

[5] Purwati P and Syahailatua A 2008 *Timun Laut Lombok barat* (Jakarta: Ikatan Sarjana Oseanologi) p 70

[6] Purwati P and Wirawati I 2009 Holothuriidae (Echinodermata, Holothuroidea, Aspidochirotida) perairan dangkal Lombok Barat, Bagian I. Genus Holothuria. *Jurnal Oseanologi* **2 (1/2)** p 1 25

[7] Purwati P and Wirawati I 2011 Holothuriidae (Echinodermata, Holothuroidea, Aspidochirotida) perairan dangkal Lombok Barat, Bagian II. Genus Actinopyga, Bohadschia, Pearsonothuria, Labidodemas. *Jurnal Oseanologi* **3 (1/2)** p 1-10

[8] Massin C 1999 Reef-dweeling Holothuroidea (Echinodermata) of the Spermonde Archipelago (South-west Sulawesi, Indonesia) *Zoologische Verhandelingen* **329** 1-144

[9] Supono, Lane D and Susetiono 2014 Echinodermata fauna of the Lembeh Strait, North Sulawesi: inventory and distribution review. *Mar. Res. Indonesia* **39 (2)** 51-61

[10] Massin C 1996 Results of the Rumphius Biohistorical Expedition to Ambon (1990). Part. 4. The Holothurioidea (Echinodermata) collected at Ambon during the Rumphius Biohistorical Expedition ed J C d Hartog *Zoologische Verhandelingen* **307** 1-53

[11] Fujita T and Marsh L M 2004 Results of the Rumphius Biohistorical Expedition to Ambon (1990).Part. 12. The Asteroidea (Echinodermata) collected from Ambon, Indonesia. *Zool. Med. Leiden* **78** 161-79

[12] Engel H 1959 *The echinoderms of Rumphius, chapter 9* in *Rumphius Memorial Volume*, ed H C D D Wit (Amsterdam: Vitgeverij en Drukkerij Hollandia N.V) p 209-21

[13] Jangoux M and Sukarno 1974 The Echinoderms Collected during The Rumphius Expedition I *Oseanologi di Indonesia* **1** 36-8

[14] Aziz A 1976 Echinoderms collected during The Rumphius Expedition II *Oseanologi di Indonesia* **6** 35-8
[15] Meyer D L 1976 The crinoidea of The Rumphius Expedition II *Oseanologi di Indonesia* **6** 39-43
[16] Aziz A1978 *Checklist of the echinoderms collections of The Rumphius Expedition II* in *International symposium on marine biogeography and evolution in the southern hemisphere* (Auckland: New Zaeland) p 717-26
[17] Aziz A 1980 Laporan ekspedisi rumphius III (3 Oktober-15 November 1977) *Oseanologi di Indonesia* **13** 27-32
[18] Setyastuti A 2009 Sea cucumber (Echinodermata:Holothurioidea: Stichopodidae, Holothuridae, Synaptidae) of West Seram, Maluku, Indonesia, collected during July 2007 *Oseanologi dan Limnologi di Indonesia* **35 (3)** 369-96
[19] Setyastuti A 2009 *Laporan Akhir Sensus Biota Laut di Teluk Ambon Proyek Insentif LIPI DIKTI: Biodiversitas Timun Laut* (Ambon: UPT BKBL Ambon LIPI) p 1-13
[20] Setyastuti A 2015 The occurence of two black sea cucumber species at Pombo Island, Central Maluku, Indonesia *Mar. Res. Indonesia* **40 (2)** 47-60
[21] Purwati P and Wirawati I 2008 Synaptidae (Echinodermata: Apodidae) dari daerah lamun El Nusa, Pulau Timor, Nusa Tenggara Timur *Oseanologi dan Limnologi di Indonesia* **34 (3)** 371-84
[22] Hellal A 2012 Taxonomic Study on the Feather Stars (Crinoidea: Echinodermata) from Egyptian Red Sea Coasts and Suez Canal, Egypt *Open Journal of Marine Science* **2 (2)** 7 DOI: 10.4236/ojms.2012.22007
[23] Hutomo M and Moosa K 2005 Indonesian marine and coastal biodiversity: present status *Indian Journal of Marine Science* **34 (1)** 88-97
[24] Sladen W 1889 *Report on the asteroidea collected by H.M.S. Challenger during the years 1873 1876. Zoology-XXX ed. Thomson C and Murray J* (London: Her Majesty's Stationary Office) p 1-893
[25] Sluiter C P 1901 Die Holothurien der Siboga Expedition. *Siboga Exped. Mon.* **44**: p. 1-142.
[26] De Meijere J 1904 *Die Echinoidea ser Siboga-Expedition. Siboga-Expeditie XLIII.*(Leiden: Buchhandlung und Druckerei) p 251
[27] Clark H 1907 *The Apodus Holothurians: A monograpgh of the synaptidae and molpadiidae.* Smithsonian contributions to knowledge, XXXV (Washington: The Smithsonian Institution) p 1-282
[28] Doderlein D 1917 *Die Asteriden Der Siboga-Expedition, I. Die Gattung Astropecten und Ihre Stammesgeschichte. Siboga-Expeditie XLVIa.*(Leiden: Boekhandel en drukkerij E.J. Brill) p 1-190
[29] Doderlein L 1920 *Die Asteriden Der Siboga-Expedition, I. Die Gattung Luidia und Ihre Stammesgeschichte. Siboga-Expeditie XLVIb* (Leiden: Boekhandel en drukkerij E.J. Brill) p 193-293
[30] Doderlein L 1924 *Die asteriden der Siboga-expedition II. Pentagonasteridae* (Leiden: Buchhandlung und druckerei vornals, E.J. Brill) p 21
[31] Doderlein L 1936 *Die Asteriden Der Siboga-Expedition, I. Die Unterfamilie Oreasterinae. Siboga-Expeditie XLVIc* (Leiden: Boekhandel en drukkerij) p 295-369
[32] Refcoll L-R 2017 *Database Voucher Echinodermata*, in *Echinodermata*, Research Center for Oceanography Reference Collection Jakarta
[33] Rumphius G E 1705 *D Amboinsche rariteitkamer, behelzende eene beschryvinge vas allerhande zoo weeke als harde schaalvischen, te weeten raare kraben, kreeften, en diergelyke zeedieren, als mede allerhande hoorntjes en schulpen, die men in dAmboinsche zee vindt: daar beneven zommige mineraalen, gesteenten, en soorten van aarde, die in d amboinsche, en zommige omleggende eilanden gevonden worden* (Amsterdam: Konftantijn den Grooten Francois Halma, Boekverkoper) p 340.

[34] Suwartana A 1980 *Notes on The Ambon research station and mollucan waters* (Stasion Penelitian Laut Ambon, Lembaga Oseanologi Nasional, Lembaga Ilmu Pengetahuan Indonesia: Ambon) p 4

[35] WoRMS 2017 *Echinodermata.* Accessed through: World Register of Marine Species at http://www.marinespecies.org/aphia.php?p=taxdetails&id=1806 [cited 2017 12-12].

[36] Paulay G 2003 The Asteroidea, Echinoidea, and Holothuroidea (Echinodermata) of the Mariana Islands *Micronesica* **35-36** 563-583

The importance of the Banda Sea for tuna conservation area: A review of studies on the biology and the ecology of tuna

W N Satrioajie[1,2], Suyadi[1,3], A Syahailatua[1] and S Wouthuyzen[4]

[1]Center for Deep-Sea Research, Indonesian Institute of Sciences (PPLD LIPI), Jl. Syaranamual Guru-Guru Poka Ambon 97233, Indonesia
[2]Aquaculture and Fisheries Group, Wageningen University, The Netherlands
[3]School of Environment, University of Auckland, New Zealand
[4]LPKSDMO, Pari Island – LIPI, Jl. Raden Saleh 43, Jakarta 10330, Indonesia

widh006@lipi.go.id

Abstract. Tuna exploitation in the Banda Sea was started when the Indonesian government gave access to the Japanese fleet under the bilateral cooperation of the Banda Sea Agreement (BSA) in 1975-1980. Several studies were conducted afterward in revealing the magnitude of tuna resources in the Banda Sea. In this paper, we reviewed the tuna studies that were done in the Banda Sea over the periods of 1980-2017 to improve our understanding on the biology and ecology of tuna in the region as the basis for future studies. Overall, we reviewed 29 publications consisted of eight research themes; biodiversity (5), catch composition (8), fish aggregating device (FAD) (2), fishing ground (2), growth-population (3), harvest-effort strategy (4), reproduction (4), and tagging (1). The Snellius II expedition in 1984-1985 was a remarkable study covering almost whole area of the Banda Sea. The study of catch composition and biodiversity were dominant in recent decade indicating there was urgent need to manage the tuna fisheries in the Banda Sea to preserve the tuna stock. Since the Banda Sea is considered as a tuna conservation area, the future studies should be focussed on the scientific findings to support the regulation.

Keywords: Banda Sea, tuna, study

1. Introduction
Indonesia is one of the main tuna producers which contributes nearly 16% or 800,000 MT of tuna per year [1]. Most of the tuna fisheries in Indonesia are operated in the eastern regions. The Banda Sea that belongs to the Indonesian fisheries management area 714 has been known as one of the most important fishing grounds for tuna [2-4]. Over 2,000 fishing vessels ranging in gears purse seine, handline, troll line, longline, and pole and line are operating in this area and catch close to 17,000 tons of tuna and tuna-like per year [5].

Historically, tuna exploitation in Indonesia has been started prior to World War II. Foreign vessel such as Japanese tuna fleet operated around the Banda Sea, the Flores Sea, the Timor Sea and the Indian Ocean close to Sumatra and Java Island [2]. After independence in 1945, Indonesia started to cease the foreign fishing activities. This act generated an objection from the Japanese government since most of the tuna industry in the country relied on the tuna resource from Indonesia.

In order to address this issue, Indonesia and Japan made an agreement which was known as the Banda Sea Agreement (BSA). During 1968-1980, there were several phases of this cooperation

Content from this work may be used under the terms of the Creative Commons Attribution 3.0 licence. Any further distribution of this work must maintain attribution to the author(s) and the title of the work, journal citation and DOI.
Published under licence by IOP Publishing Ltd

IOP Conf. Series: Earth and Environmental Science **184** (2018) 012004 doi:10.1088/1755-1315/184/1/012004

negotiating the fisheries access of the Japanese vessels to operate in the Indonesian waters [2]. Indonesia's goals were to earn foreign exchange, income, employment, technology transfer and regional development. These missions were considered to be very important during that period since Indonesia needed additional income immediately to develop its regions after the independence. On the Japanese side, the BSA gave the benefit in maintaining protein consumptions in the country through tuna supply from Indonesia. This arrangement brought about 40,000 MT tuna from the Banda Sea to Japan. During the same period, Indonesia earned about US$20 million to support the state budget. Indonesia expanded this policy by giving fishing licenses to foreign vessels from other countries even though this latter decision recently has been temporarily suspended under Ministerial Regulation of Marine Affairs and Fisheries (MAF) 56 2014 since many foreign vessels were indicated to be main actors of illegal unreported unregulated (IUU) fishing [6].

Concerning the importance of marine resources, in terms of history of marine research, there were six expeditions held in the colonial era where the Banda Sea was one of the area of interests (Table 1). Although those expeditions were more focused on the deep-sea biota and oceanographic features, the scientific findings were significant in revealing natural resources of the Banda Sea [7, 8]. There were seven expeditions conducted during five decades after the independence (Table 1) [7, 9]. Snellius II (1984-1985) was the biggest expedition ever carried out under the Indonesia-Netherlands cooperation where more than 450 scientists and technicians joined this research. This expedition covered most areas of the Banda Sea [7]. Furthermore, for the first time, tuna that belongs to the family of *Scombridae* have been found explicitly in some parts of the reports of this research. Apart from the rapid development of tuna fisheries on that period [2], this result has become a starting point concerning the importance of tuna biology and ecology studies in the Banda Sea [10, 11].

Table 1. Expeditions occurred in the Banda Sea [7, 9].

No	Expedition	Year
1	Challenger	1872-1876
2	Gazelle	1875
3	Siboga	1899-1900
4	Mortensen	1922
5	Vening-Meinesz	1927-1929
6	Snellius I	1929-1930
7	Galathea	1950-1952
8	Baruna I	1964
9	Alpha Helix	1975 and 1979
10	SEATAR	1966-1980
11	Corindon	1980-1981
12	Snellius II	1984-1985
13	Karubar	1992
14	Baruna VII	2013

Given the increase of fishing effort in the Banda Sea, Indonesian government recently has considered the exploitation level of tuna in the Banda Sea as fully-exploited [5]. A specific regulation which prohibited fishing activities in a huge closure area in the Banda Sea (126-132°E, 4-6°S) or almost 130,000 km^2 (20%) of the total area during the period of October – December each year was stated under MAF Ministerial Regulation No 4 2015 (Figure 1). This area was stipulated to be a spawning ground of yellowfin tuna *(Thunnus albacares)* that is one of the important tuna commodities in Indonesia [1]. Nevertheless, this regulation leads to a critic from the stakeholders that question the scientific basis of the closure area in the Banda Sea.

Some previous studies have discussed the potential impacts of this regulation on socio-economic [12] and fishing competition to other areas [6]. However, there is still a lack of information that supports the idea of spawning ground of tuna existence in the Banda Sea. Indeed, the number of tuna studies within the focus area in the Banda Sea increased after the 1990s. However, the scientific findings of those tuna studies are highly fragmented.

Reviewing tuna studies in the Banda Sea is needed to obtain better understanding on the current status of tuna stock and the potential issues. This paper reviews the scientific literature on past tuna studies and compiles empirical proofs in revealing the tuna resources in the Banda Sea. The finding of each study will be used to trace the idea of tuna spawning ground occurred in the Banda Sea and also to provide the scientific basis of closure area in the Banda Sea. In addition, we propose future studies that could be applied regarding tuna biology and ecology, to strengthen the aforementioned regulation.

2. Literature Review

We collected information of original publications of tuna studies in the Banda Sea from the last three decades. Information was derived by searching in Scopus, ISI Web of Science (Thomson Reuters), and Google Scholars using query combination "tuna banda sea" or "banda sea tuna". Some local publications, graduate theses, and reports were also used to enhance the material including the proceedings.

As it can be expected that many research on fisheries and or tuna have been done in the Banda Sea, only publications focusing on the biology and ecology of tuna were selected in this paper. From those topics, only publications that contained subject(s) such as; 1) fishing ground, 2) biodiversity, 3) reproduction, 4) catch composition, 5) fish aggregating device (FAD), 6) growth population, 7) tagging-movement and 8) harvest-effort strategy were included in the review. Overall, there were 29 publications collected where each publication, the main species, study area, observation were recorded (Table 2). Publications of the Snellius II expedition in 1990 were used as the oldest literature since the pelagic fish study was first done in this research [4, 13].

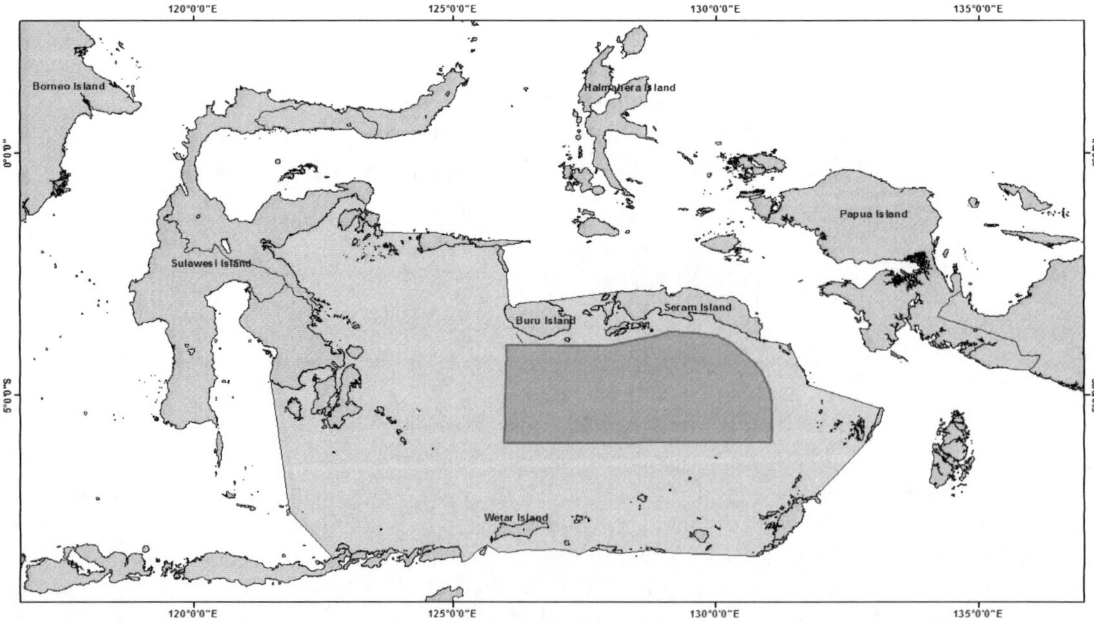

Figure 1. Study area of the Banda Sea belongs to Fisheries Management Area 714 within blue colour. The closure area is inside the red line based on Ministerial MAF 4 2015.

Table 2. Summary of tuna studies in the Banda Sea.

No	Observation	Subject	Methodology	Species	Area	Ref.
1	Fish abundance was 75% higher in the depth layer 3-100 meters during night observation.	Biodiversity	Acoustic	Pelagic fish	Whole Banda Sea	[4]
2	Fish larvae *Scombridae* was found in depth 0-100 meters during both day and night sampling.	Biodiversity	Midwater trawl	Fish larvae	Whole Banda Sea	[13]
3	Mollusks were dominant in stomach content and SKJ spawned in the whole year within peak season in June.	Reproduction	Stomach and Gonad observation	SKJ	North Banda close to South Seram Island	[14]
4	*Scombridae* was found the second highest within 26% of total larvae.	Biodiversity	Bongo net	Fish larvae	Middle Banda Sea close to Banda Islands	[15]
5	Spawning season was estimated to occur in December.	Reproduction	Onboard observer	BET	Southwest Banda Sea close to Wetar and Alor	[16]
6	Bait fish were found dominant in the stomach content and the gonad maturity did not link with fish length and age.	Reproduction	Stomach and Gonad observation	SKJ	South Seram and Nusa Laut Island	[17]
7	Range of fork length (FL) caught by longline was 104-123cm for yellowfin and 104-153cm for big eye.	Catch composition LL	Onboard observer	BET YFT	Close to Banda Islands 5-6°S and 129-130°E	[18]
8	Catch of pole and line was dominated by SKJ (FL 30-78cm) and followed by juvenile of yellowfin and big eye tuna.	Catch composition PL	Onboard observer	SKJ, YFT and BET	North Banda, Ambon and Buru Island	[19]
9	Mini purse seine caught 30-36% of bullet tuna.	Catch composition PS	Onboard observer	Bullet tunas *Auxis* spp	Close to Kendari	[20]
10	High abundance were found near Hatta and Gunung Api Island by 49 individual/liter and 37 individual./liter respectively.	Biodiversity	Acoustic and bongo net	Tuna larvae	North Banda Sea from Ambalau to Banda Neira	[21]
11	Tuna production was estimated 123.9-162.4kg/trip with peak season in October-December.	Harvest-effort strategy	Interview survey	Tuna	Banda Neira Island	[22]
12	Range of fork length (FL) modus caught by purse seine was 44-46cm, pole and line 40-42cm, handline 38-40cm and troll line 44-46cm.	Catch composition PS, PL, TL and HL	Port data	SKJ	Kendari	[23]
13	There were 200 fish tagged in the Banda Sea of total number 25,197 fish. FL of tagged fish were 28-65cm of yellowfin, 29-83 of SKJ and 29-58 of big eye.	Tagging-migration	Onboard observer	YFT, SKJ and BET	North Banda close to Ambon and Saparua	[24]

continued on the next page

IOP Conf. Series: Earth and Environmental Science **184** (2018) 012004 doi:10.1088/1755-1315/184/1/012004

Table 2. Continued

No	Observation	Subject	Methodology	Species	Area	Ref.
14	Range of fork length (FL) was 55-215cm and natural and fishing mortality rate were 2.4yr^{-1} and 1.79yr^{-1} respectively.	Growth-population	Enumeration	YFT	East Banda Sea 4-6°S and 129-130.30°E	[25]
15	There was no genetic differentiation between yellowfin caught in Ambon and North Molucca.	Biodiversity	Genetic	YFT	Ambon	[26]
16	Range of fork length (FL) of yellowfin was 87.5-177.5cm and the modus of size was 137.5cm.	Catch composition HL	Onboard observer	YFT	Banda Neira Island	[27]
17	The size (FL) of SKJ decreased over three decades 45-50.9cm (1984-1986), 42.4-48.5cm (2007-2011).	Catch composition PL	Onboard and enumeration	SKJ	Ambon, Lease Island, South Buru	[28]
18	Length and weight of SKJ caught unassociated with FADs was higher than FADs associated.	Catch composition PL	Onboard observer	SKJ	West Banda Sea close to Wawoni and Buton	[29]
19	Range of fork length (FL) caught by troll line was 20-70cm with modus 40-50cm.	Catch composition TL	Onboard observer	SKJ YFT	Kendari	[30]
20	There were 39 FADs recorded on the survey.	FAD	Interview survey	Tuna	Ambon	[31]
21	The total number of FADs in the period of 1997-2002 in the Banda Sea based on government permission was 27.	FAD	Port data	Tuna	Banda Sea (not specific)	[32]
22	The highest probability of SKJ fishing ground was estimated located in West Seram, Buano, Kelang and Manipa.	Fishing ground	Remote sensing and logbook	SKJ	Banda Sea 2-5°Sand 125-131.5°E	[33]
23	The criteria of size catch was >58cm.	Growth-population	Enumeration	SKJ	Ambon	[33]
24	MSY of SKJ was 33.000ton/year.	Harvest-effort strategy	Interview survey and Port data	SKJ	Banda Sea (not specific)	[34]
25	Yellowfin spawned all year within peak season in October – December.	Reproduction	Enumeration	YFT	Around Banda Islands 3-4°S and 129-130°E	[35]
26	By catch of tuna longline was dominated by sickle pomfret, stingray and lancet fish.	Catch composition LL	Onboard observer	Tuna	Southwest Banda Sea close to Wetar and Alor	[36]
27	Fishing ground of SKJ was predicted in east Wawonii Island.	Fishing ground	Remote sensing	SKJ	West Banda Sea close to Kendari	[37]
28	Range of fork length (FL) of SKJ was 36.54-51.30cm and maximum length ($L\infty$) was 70.1cm	Growth-population	Enumeration	SKJ	Kendari	[38]
29	Maximum Sustainable Yield of SKJ was 180.487ton and total allowable catch was 144.390ton.	Harvest-effort strategy	Vessel logbook and port data	SKJ	North Banda Sea close to Ambon	[39]

HL = Handline, PL= Pole and line, TL= Troll line, LL= Longline, and PS= Purse Seine, SKJ= Skipjack Tuna, YFT= Yellowfin Tuna, BET= Big Eye Tuna, MSY= Maximum Sustainable Yield

3. Development of the tuna studies in the Banda Sea

In general, the number of the tuna studies in the Banda Sea increased significantly since the 1990s (Figure 2). Despite the Snellius II expedition was not specifically purposed for tuna study, it was a milestone in revealing the magnitude of marine resources including tuna in the Banda Sea. The expedition encompassed almost the whole area of the Banda Sea involving multidisciplinary of oceanography research: geological, physical, chemical, and biological [9]. Given the large area of the Banda Sea, there were only two studies that discussed the small size of potential tuna fishing grounds based on satellite imagery. They suggested that the fishing ground located close to the islands covering Ambon and Buru Islands in the north Banda Sea and Wawonii and Wakatobi Islands in the west Banda Sea [33, 37]. Two important commodities namely the skipjack (SKJ) / *Katsuwonus pelamis* and the yellowfin tuna (YFT) / *Thunnus albacares* were the most frequent subjects of the studies. There was only small number of studies concerning ~~on~~ the big eye (BE) *Thunnus obesus* and other tuna-like species such as *Auxis* sp (Table 2).

There were two publications from the Snellius II expedition related to tuna focusing on the abundance of pelagic fish including tuna and its larvae [4, 13]. Those studies were updated by Taufik et al. in 2005 and Karsono et al. in 2012, reporting the high abundance of tuna larvae around Banda Islands and Ambalau Island respectively. Those studies suggested that most tuna larvae were found in the depth layer of 0-100meters [4, 13, 15]. Meanwhile, tuna reproduction study was first studied in the first decade focussing on skipjack [14]. It was followed by another study discussing on other species BE and YFT. Those studies found that YFT and SKJ spawn throughout the year, within peak seasons in October to December [14, 16, 17, 35].

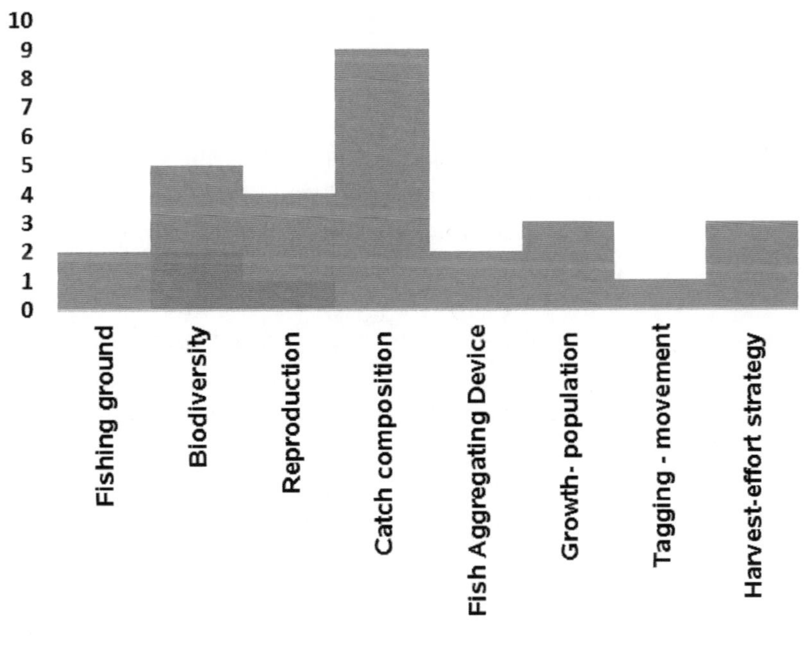

Figure 2. Computation of selected tuna subject publications in the Banda Sea in three decades.

More studies were focused on catch composition from different fishing gears and harvest control rules in the last two decades [27-30]. This indicated more concerns in revealing the fishing impact towards tuna stock [40]. Catch composition of five main tuna gears that operated in the Banda Sea was observed. Handline and longline vessels caught bigger size of YFT than troll line [18, 30]. The SKJ was the main species of pole and line fishery operated in this area [19, 28]. Some studies observed small purse seine fishery reporting that SKJ, tuna-like species such as *Auxis* spp and small pelagics fish were the dominant species. However, there were also concerns about the decrease in the size of SKJ and YFT and also the potential catching of juvenile tuna on several gears such as purse seine, longline, pole and line [18, 23, 28]. These findings were also encountered by some studies discussing the fish growth and fishing mortality of SKJ and YFT revealing that those species were caught in immature conditions [25, 33, 38].

Tuna fishery in the Banda Sea cannot be separated also from the use of fish aggregating devices (FADs). This device has been a concern on tuna sustainability of Indonesian tuna fisheries since the 2000s [41, 42]. A study in western Banda Sea reported that smaller size of SKJ was caught around this device [29]. Some descriptive studies also already reported the total number of FADs on the basis of licenses and interview survey, however, the distribution and density of FADs in this Banda Sea remain unclear [31, 32]. With regards to tuna movement, the use of archival tags to determine the migration patterns was studied in eastern Indonesia [41]. Unfortunately, there was only one paper discussing the number of fish that were tagged in the north of the Banda Sea close to the Ambon Island. Meanwhile, on the stock study, SKJ that associated with pole and fishery was used as the case study to estimate the maximum sustainable yield. From both studies, it was found that the exploitation level of skipjack around Ambon was under fished and it was suggested that pole and line fishery could be maintained [33, 39].

4. Idea of tuna spawning ground and closure area

The investigation of tuna spawning location has been approached using several variables such as larvae abundance [43-45], tuna behaviour and movement patterns [46-48], and interaction between larvae and oceanography feature such as temperature [49]. Following that schemes, several tuna larva studies have been done in the Banda Sea revealing that this larva was abundant [13, 15, 21]. However, there was no clear record that the spawning ground of tuna, particularly YFT existed in the Banda Sea. Those studies only took snapshots of the tuna larvae abundance in a short period from specific locations making it really difficult to describe the role of the Banda Sea related to the life cycle of tuna. Despite the closure period in October until December is according to the reproduction study of YFT [35], more studies are needed to sharpen the analysis, by expanding the observation from another area surrounding the Banda Sea.

The realistic reason in justifying the importance of closure area of the Banda Sea is the harvest-effort and catch composition studies. However, it should be noticed that the national tuna management plan (RPP-TCT) under Ministerial Decree MAF No. 107 2015 also did not propose a closure area as a part of the strategy in sustaining tuna stock. This regulation emphasized to control (hold) the fishing effort in the Banda Sea despite the exploitation level that was still under MSY. This concern is in line with some literature concerned on the decrease of the length of fish caught from different fishing gears and catch amount on the basis of observer and enumeration program [19, 23, 28].

5. Implication on the management and future studies

All studies showed concerns regarding the depletion of tuna resources in the Banda Sea. By contrast, we argue that the regulation of the restriction area under the Ministerial Regulation of MAF 4 2015 in the Banda Sea has a strong scientific background. Some studies indicated the finding of tuna larvae in some area of the Banda Sea. However, it is noteworthy that no study has reported tuna post-larvae or early tuna juveniles exist in some periods which suggested that this area could be considered as a conservation area for specific species.

In general, the idea of making a closure area to promote the recovery of YFT stock could potentially be beneficial. However, the decision of location choices, sizes, shapes, and closure duration (seasonal or multi-annual duration) should be carefully studied. Furthermore, the relevancy with other regulations such as Ministerial Decree MAF 69 2009 that organized the marine conservation area and Ministerial Decree MAF 58 2014 that managed the establishment of aquatic tourism park in the Banda Islands of the area should be considered carefully. In particular, knowledge of where the spawning and nursery sites are located and how these locations vary with population characteristics (abundance, age/size structure) remains highly needed.

This paper is the first to review the tuna research in the Banda Sea. Given the difficulties in obtaining reliable findings from the various studies, it can provide the urgently required insight as compiling the snapshots to the better quality of the scientific advice in managing tuna resources in the Banda Sea. We suggest the future studies should be prioritized on exploring the accuracy of regulation such as investigating the life cycle of the two main species SKJ and YFT in Indonesian waters either regularly larvae survey, advance tagging study, and genomics. Another urgent study needs to be done is to evaluate the implementation of Ministerial Regulation of MAF 4 2015 and the impact of the closure area to other aspects such as production and socio-economy.

References

[1] Sunoko R and Huang H-W 2014 Indonesia tuna fisheries development and future strategy *Mar. Policy* **43** 174-83

[2] Marten G, Matsuda Y, Bardach J, Comitini S and Hardjolukito S 1981 A Strategic goal analysis of options for tuna longline joint ventures in Southeast Asia: Indonesia-Japan case study *Research Report No.3 East-West Environmental and Policy Institute* p 73

[3] Matsuda Y 1987 Postwar development and expansion of Japan's tuna fishery Tuna Issues and Perspectives in the Pacific Islands Region *Tuna issues and perspectives in the Pacific islands region* eds Doulman D J East-West Center, Hawai'i 71-91

[4] Amin E M and Nugroho D 1990 Acoustic surveys of pelagic fish resources in the Banda Sea during August 1984 and February–March 1985 *Neth. J. Sea Res.* **25**(4) 621-6

[5] Suman A, Wudianto, Sumiono B, Irianto H E, Badrudin and Amri K 2014 *Potensi Dan Tingkat Pemanfaatan Sumberdaya Ikan di Wilayah Pengelolaan Perikanan Republik Indonesia (WPP RI)* (Jakarta: Ref Graphika and Balai Penelitian Perikanan Laut Pusat Penelitian Pengelolaan Perikanan dan Konservasi Sumber Daya Ikan Badan Penelitian dan Pengembangan Kelautan dan Perikanan) p 197

[6] Duggan D E and Kochen M 2016 Small in scale but big in potential: Opportunities and challenges for fisheries certification of Indonesian small-scale tuna fisheries *Mar Policy* **67** 30-9

[7] Nontji A 2009 *Penjelajahan dan penelitian laut Nusantara dari masa ke masa* Jakarta Pusat Penelitian Oseanografi-LIPI p 433

[8] Sumadhiharga O K 2009 *Ikan tuna* Pusat Penelitian Oseanografi-LIPI p 129

[9] Nontji A 2005 *Laut Nusantara* 4th revised edition (Jakarta: Djambatan) p 372

[10] Dalzell P and Pauly D 1990 Assessment of the fish resources of southeast Asia, with emphasis of the Banda and Arafura seas *Neth. J. Sea Res.* **25**(4) 641-50

[11] Butcher J G 2004 *The closing of the frontier: A history of the marine fisheries of Southeast Asia*, c 1850-2000: Institute of Southeast Asian Studies p 442

[12] Adam L 2016 Kebijakan Pelarangan Penangkapan Ikan Tuna Sirip Kuning: Analisis Dampak dan Solusinya *Jurnal Ekonomi dan Kebijakan Publik* **7**(2) 215-27

[13] Soewito and Schalk P H 1990 Spatial and seasonal patterns in fish larvae distribution in the Banda Sea (Indonesia) *Neth. J. Sea Res.* **25**(4) 591-600

[14] Wouthuyzen S, Peristiwady T, Manik N, Djoko D, and Hukom F 1990 Makanan dan aspek reproduksi ikan cakalang *(Katsuwonus pelamis)* di Laut Banda, suatu studi perbandingan *Perairan Maluku dan Sekitarnya* 1-16

[15] Taufik M, Suwarso S and Nurwiyanto N 2005 Distribusi Kelimpahan Ichthyoplankton di Teluk Tomini dan Laut Banda *Jurnal Penelitian Perikanan Indonesia* **11**(6) 73-83

[16] Nugraha B and Mardlijah S 2006 Hubungan Panjang Bobot, Perbandingan Jenis Kelamin dan Tingkat Kematangan Gonad Tuna Mata Besar *(Thunnus obesus)* di Perairan Laut Banda *Jurnal Penelitian Perikanan Indonesia* **12**(3) 195-202

[17] Manik N 2007 Beberapa aspek biologi ikan cakalang *(Katsuwonus pelamis)* di perairan sekitar Pulau Seram Selatan dan Pulau Nusa Laut *Oseanologi dan Limnologi di Indonesia* **33** 17-25

[18] Nugraha B and Chodrijah U 2010 Komposisi Hasil Tangkapan dan Daerah Penangkapan Kapal Tuna Longline di Perairan Laut Banda *Jurnal Penelitian Perikanan Indonesia* **16**(4) 305-9

[19] Nugraha B, Mardlijah S and Rahmat E 2010 Komposisi Ukuran Cakalang *(Katsuwonus pelamis)* Hasil Tangkapan Huhate yang Didaratkan di Tulehu, Ambon *BAWAL Widya Riset Perikanan Tangkap* **3**(3) 199-207

[20] Hariati T 2011 Komposisi hasil tangkapan, musim penangkapan, dan indeks kelimpahan ikan pelagis yang tertangkap pukat cincin mini di perairan Kendari, Laut Banda *Jurnal Penelitian Perikanan Indonesia* **17**(2) 139-46

[21] Wagiyo K, Chodrijah U and Hany Y 2012 Kelimpahan dan Sebaran Larva *Scombridae* di Perairan Laut Banda *Status Pemanfaatan Sumber Daya Ikan di Perairan Selat Makassar, Teluk Bone, Laut Flores dan Laut Banda* eds Suman A, Wudianto and Sumiono B (Bogor: Bogor Agricultural University - IPB Press) p 300

[22] Baihaqi and Yusuf H N 2012 Aktivitas Penangkapan Pancing Ulur Tuna di Kepulauan Banda Neira *Status Pemanfaatan Sumber Daya Ikan di Perairan Selat Makassar, Teluk Bone, Laut Flores dan Laut Banda* eds Suman A, Wudianto and Sumiono B (Bogor: Bogor Agricultural University - IPB Press) p 300

[23] Noegroho T, Hidayat T and Chodrijah U 2012 Struktur Ukuran dan Daerah Penangkapan Ikan Cakalang *(Katsuwonus pelamis)* yang Berbasis di PPS Kendari *Status Pemanfaatan Sumber Daya Ikan di Perairan Selat Makassar, Teluk Bone, Laut Flores dan Laut Banda* eds Suman A, Wudianto and Sumiono B (Bogor: Bogor Agricultural University - IPB Press) p 300

[24] Natsir M, Widodo A A and Prisantoso B I 2012 Fishing Activity, Size Distribution, Tag Released and Recapture of Tuna Tagging in Eastern Indonesian Waters *Indonesian Fisheries Research Journal* **18**(1) 47-56

[25] Damora A and Baihaqi 2013 Struktur Ukuran Ikan dan Parameter Populasi Madidihang *(Thunnus albacares)* di Perairan Laut Banda *BAWAL Widya Riset Perikanan Tangkap* **5**(1) 59-65

[26] Akbar N 2014 Keragaman Genetik, Struktur Populasi dan Filogenetik Ikan Tuna Sirip Kuning *(Thunnus albacares)* di Perairan Maluku Utara dan Ambon, Indonesia (Bogor: Bogor Agricultural University) p 44

[27] Chodrijah U 2015 Hubungan Panjang Bobot dan Struktur Ukuran Ikan Madidihang *(Thunnus albacares)* di Perairan Laut Banda *Proc. Simposium Pengelolaan Perikanan Tuna Berkelanjutan* 10-11 December 2014 Bali p 1238

[28] Syahailatua A and Pay L 2015 Struktur Ukuran Ikan Cakalang (*Katsuwonus Pelamis*) di Perairan Ambon dan Implikasinya Bagi Pengelolaan *Proc Simposium Pengelolaan Tuna Berkelanjutan* 10-11 December 2014 Bali p 1238

[29] Husair, Tadjuddah M, Abdullah, Anadi L, Mustafa A, Arami H, 2015 Analisis Hasil Tangkapan Ikan Cakalang *(Katsuwonus pelamis)* Pada Daerah Penangkapan dengan Menggunakan Rumpon dan Tanpa Rumpon di Perairan Barat Laut Banda *Proc. Simposium Nasional Pengelolaan Perikanan Tuna Berkelanjutan* 10-11 December 2014 Bali p 1238

[30] Hidayat T, Chodrijah U, and Noegroho T 2014 Karakteristik Perikanan Pancing Tonda di Laut Banda *Jurnal Penelitian Perikanan Indonesia* **20**(1) 43-51

[31] Hargiyatno I T, Anggawangsa R F, Samusamu A S, and Widodo A A 2015 Distribusi dan Jarak Pemasangan Rumpon Laut Dalam Dalam Upaya Pengelolaan Perikanan Tuna yang Berkelanjutan (Studi Kasus di Kendari, Maumere, Ambon dan Pelabuhan Ratu) *Proc. Simposium Nasional Pengelolaan Perikanan Tuna Berkelanjutan* 10-11 December 2014 Bali p 1238

[32] Nasution A M 2014 Kebijakan Penataan Rumpon dan Armada Pukat Cincin di Indonesia *Proc.Simposium Pengelolaan Perikanan Tuna Berkelanjutan* 10-11 December 2014 Bali p 1238

[33] Waileruny W, Wiyono E S, Wisodo S H, Purbayanto A and Nurani T W 2014 Monsson and Skipjack Fishing Ground in The Banda Sea and Its Surrounding Moluccas Province *Jurnal Teknologi Perikanan dan Kelautan* **5**(1) 41-54

[34] Waileruny W, Wiyono E S, Wisudo S H, Nuraini T W and Purbayanto A 2015 Model dinamis pemanfaatan berkelanjutan sumberdaya perikanan cakalang di Laut Banda dan Sekitarnya Provinsi Maluku *Proc. Simposium Pengelolaan Perikanan Tuna Berkelanjutan* 10-11 December 2014 Bali p 1238

[35] Wagiyo K, Suman A and Patria M P 2015 Sebaran dan Hubungan Parameter Reproduksi Ikan Tuna Madidihang *(Thunnus albacares)* dengan Suhu dan Klorofil-a di Laut Banda *BAWAL Widya Riset Perikanan Tangkap* **7**(3) 183-91

[36] Nugraha B and Wagiyo K 2006 Hasil tangkapan sampingan (by-catch) tuna long line di perairan Laut Banda *BAWAL Widya Riset Perikanan Tangkap* **1**(2) 71-5

[37] Agusliana S E, Tadjuddah M and Mustafa A 2017 Sebaran suhu permukaan laut dan tracking daerah penangkapan Ikan Cakalang di Perairan Barat Laut Banda *Jurnal Manajemen Sumber Daya Perairan* **2**(1) 41-9

[38] Tadjuddah M, Anadi L, Mustafa A, Arami H, Kamri S and Wianti N I 2017 Growth pattern and size structure of skipjack tuna caught in Banda Sea, Indonesia *Aquaculture, Aquarium, Conservation & Legislation-International Journal of the Bioflux Society (AACL Bioflux)* **10**(2) 227-33

[39] Soukotta I V, Bambang A N, Sya'rani L and Saputra S W 2017 Estimation of MSY and MEY of skipjack tuna *(Katsuwonus pelamis)* fisheries of Banda Sea, Moluccas *Aquaculture, Aquarium, Conservation & Legislation-International Journal of the Bioflux Society (AACL Bioflux)* 2017 **10**(2) 435-44

[40] Suman A, Irianto H E, Satria F and Amri K 2016 Potensi dan Tingkat Pemanfaatan Sumberdaya Ikan di Wilayah Pengelolaan Perikanan Negara Republik Indonsia (WPP NRI) Tahun 2015 serta Opsi Pengelolaannya *Jurnal Kebijakan Perikanan Indonesia* **8**(2) 97-100

[41] Natsir M and Proctor C 2011 An overview of FAD-based fisheries and FAD issues in Indonesian Archipelagic waters *Proc. Tuna Fisheries and Fish Aggregating Devices* 29 November Tahiti 2011

[42] Satrioajie W N, de Froe E, van Zwieten P A M and Rijnsdorp A D 2014 Effort Allocation of Anchored Fish Aggregation Devices (FADs)- Based Fisheries in the North Sulawesi, Indonesia *Proc. The 6th International Conference on Agribusiness Economics and Management (ICAEM) 2014*; 2 September 2014 Davao (Davao: University of the Philippines Mindanao Mintai, Davao) pp 6-12

[43] Richardson D E, Marancik K E, Guyon J R, Lutcavage M E, Galuardi B and Lam C H 2016 Discovery of a spawning ground reveals diverse migration strategies in Atlantic bluefin tuna *(Thunnus thynnus) Proc. the National Academy of Sciences* **113**(12) 3299-304

[44] Alemany F, Quintanilla L, Velez-Belchí P, García A, Cortés D and Rodríguez J M, 2010 Characterization of the spawning habitat of Atlantic bluefin tuna and related species in the Balearic Sea (western Mediterranean) *Progress in Oceanography* **86**(1) 21-38

[45] Mariani P, MacKenzie B R, Iudicone D and Bozec A 2010 Modelling retention and dispersion mechanisms of bluefin tuna eggs and larvae in the northwest Mediterranean Sea *Progress in Oceanography* **86**(1) 45-58

[46] Teo S L, Boustany A, Dewar H, Stokesbury M J, Weng K C and Beemer S 2007 Annual migrations, diving behavior, and thermal biology of Atlantic bluefin tuna, *Thunnus thynnus*, on their Gulf of Mexico breeding grounds *Mar. Biol.* **151**(1) 1-18

[47] Teo S L, Boustany A M and Block B A 2007 Oceanographic preferences of Atlantic bluefin tuna, *Thunnus thynnus*, on their Gulf of Mexico breeding grounds *Mar. Biol.* **152**(5) 1105-19

[48] Aranda G, Abascal F J, Varela J L and Medina 2013 A spawning behaviour and post-spawning migration patterns of Atlantic bluefin tuna (*Thunnus thynnus*) ascertained from satellite archival tags *Plos One* **8**(10) e76445

[49] Lehodey P, Senina I and Murtugudde R 2008 A spatial ecosystem and populations dynamics model (SEAPODYM)–Modeling of tuna and tuna-like populations *Progress in Oceanography* **78**(4) 304-18

Geological aspects of Banda Sea ecosystems and how they shape the oceanographical profile

J M Pownall[1], R Hall[2], G S Lister[1] and A Trihatmojo[3]

[1]Research School of Earth Sciences, Australian National University, Canberra, ACT 2601, Australia
[2]SE Asia Research Group, Department of Earth Sciences, Royal Holloway University of London, Egham TW20 0EX, United Kingdom
[3]Fakultas Ilmu dan Teknologi Kebumian, Institut Teknologi Bandung, Jalan Ganesha 10, Bandung 40132, Indonesia

E-mail: jonathan.pownall@anu.edu.au

Abstract. The Banda Sea is a collage of young oceanic basins and fragmented Australian continental crust located at the heart of the Australia–SE Asia collision zone where Australian and Asian biogeographic regions converge. The formation of the sea was governed by the southeastward rollback of the Banda Slab since *c.* 16 Ma, which in its wake opened new oceanic basins and extended and fragmented Australian crust. These Australian crustal fragments are today either stranded within the Banda Sea where they form the prominent submarine 'Banda Ridges', or now reside as thrust-sheets on the NW Australian shelf after being transported all the way to the southern Banda Arc. The deepest part of the Banda Sea, the 7.2 km Weber Deep, was formed by extreme lithospheric extension that occured in the latter stages of Banda Slab rollback. This extension was accommodated by the vast low-angle 'Banda Detachment', which operated above the subducted fringes of the Australian continental margin.

1. Introduction

The Banda Sea is one of several seas located within the Indonesian archipelago between Australia and mainland SE Asia. Together, these seas comprise the route for the Indonesian 'Throughflow' between the Indian and Pacific oceans – a crucial gateway for ocean circulation and regulation of global climate [1, 2, 3]. These seas also mark the boundary for important biological separations. Wallace's line passes through the Celebes Sea and west of Sulawesi, while Weber's line and Lydekker's line both pass through the Banda Sea its self. These observations can all be explained when considering the region's changing tectonics. The formation of narrow seaways, and the juxtaposition of different flora and fauna, are both side-effects of the collision of Australia with SE Asia and the closure of the Tethys Ocean that once separated them [4] (Fig. 1).

One hundred and eighty million years ago (Ma), Australia and India both separated from Antarctica during breakup of the supercontinent Gondwana, later heading north. India collided with central Asia at around 50 Ma, ahead of the slower continent of Australia that collided with SE Asia at around 23 Ma [5]. Whereas India–Asia collision commenced early enough to have today produced the Himalaya, Australia–Asia collision is still in its infancy. In the place of 8000-metre peaks and an elevated continental plateau is a complex array of oceanic basins, continental fragments, volcanic arcs, and carbonate platforms. The infancy of the Australia–SE Asia collision zone accounts for why the geography and ecosystems of the Indonesian archipelago and its intervening seaways are so intricate and diverse.

Content from this work may be used under the terms of the Creative Commons Attribution 3.0 licence. Any further distribution of this work must maintain attribution to the author(s) and the title of the work, journal citation and DOI.
Published under licence by IOP Publishing Ltd

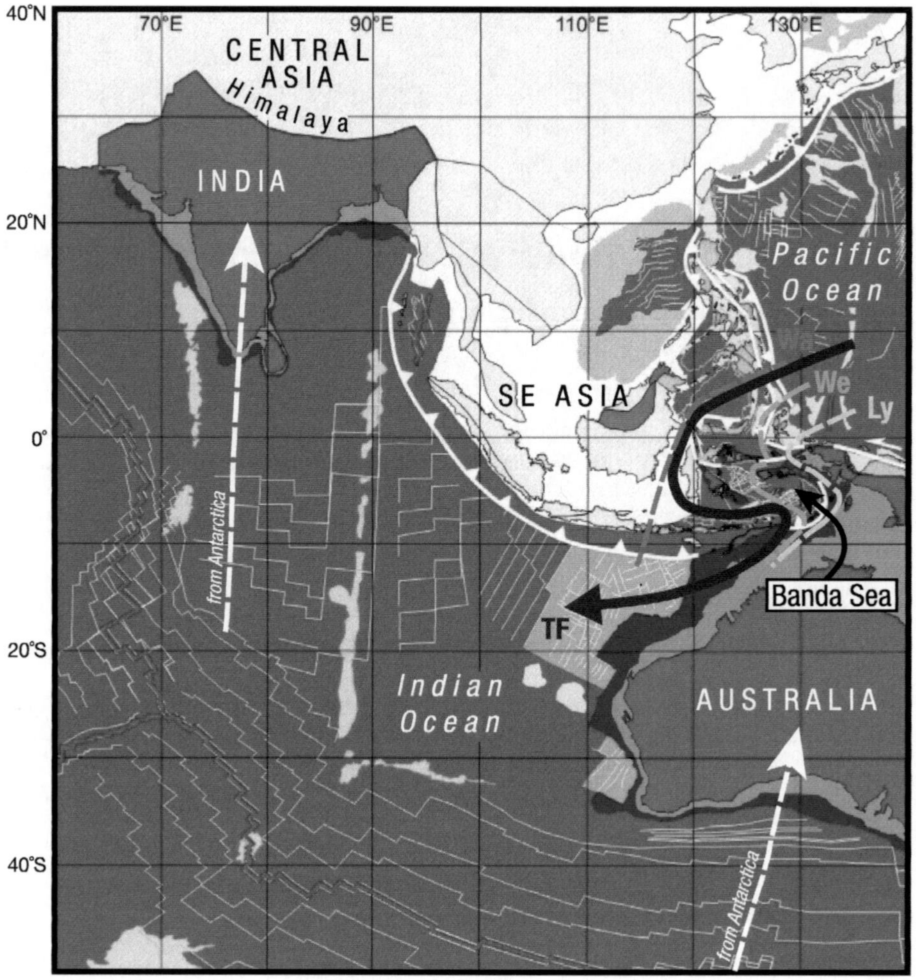

Figure 1. Map of India, SE Asia, and Australia. The Banda Sea is located within the SE Asia–Australia collision zone. India and Australia (Gondwanan; red), which both rifted from Antarctica at *c.* 160 Ma, collided with Asia (yellow) at *c.* 50 Ma and *c.* 23 Ma, respectively. Note the course of the Indonesian Troughflow (TF) oceanic current connecting the Indian and Pacific oceans, and the positions of Wallace's (Wa), Weber's (We), and Lydekker's (Ly) faunal boundary lines within the SE Asia–Australian collision zone. Figure adapted from Hall (2012) [4].

2. How did the Banda Sea form?

The Banda Sea has a complex and varied oceanographic profile due to the composite nature of the crust it overlies [6]. Young oceanic crust (12.5–3.5 Ma [7, 8]) contains numerous continental fragments such as those forming the Banda Ridges, and features the very deep oceanic basins of the North Banda Basin and Weber Deep (Fig. 2). The Banda Sea lies mainly within the curved island chain of the Banda Arc, from Timor to Tanimbar to Kai to Seram to Buru. An active inner volcanic arc from Damar volcano to Banda Api [9, 10] forms an eastward extension to the more mature volcanic islands of Flores, Alor, and Wetar. The 7.2 km deep Weber Deep [11, 12] is the forearc basin between the inner volcanic and outer non-volcanic arcs.

So, why is the Banda Arc so tightly curved? It's shape was dicated by the geometry of the underlying subducted slab, shown by the location of earthquakes produced within it to have assumed a highly

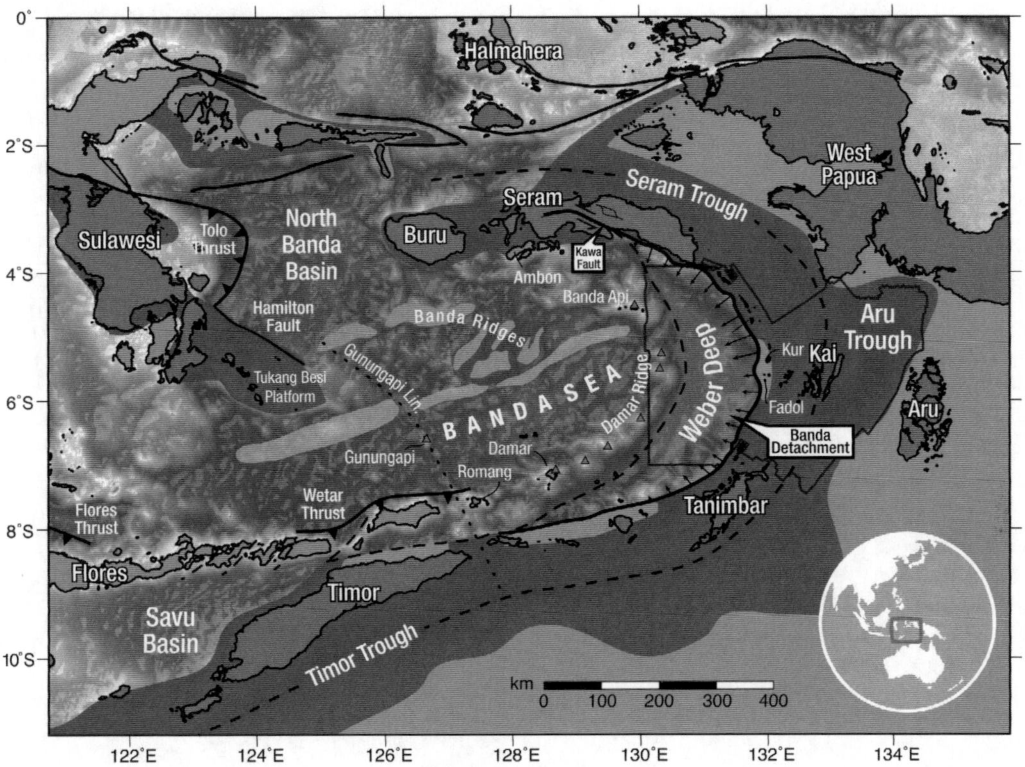

Figure 2. Present-day configuration of the Banda Sea, showing the location of the Banda Detachment flooring the Weber Deep [11]. As in Fig. 4, submarine arcs and oceanic plateaus are shown in pale blue; volcanic island arcs, ophiolites and material accreted along plate margins are shown in green; and Australian-affinity continental crust is shown in red.

concave spoon- or half-bathtub-shaped geometry [13, 14, 15] (Fig. 3). This slab, an eastward extension to the Java and Sumatra slabs, was once located adjacent to southern Sulawesi – a few thousand kilometres further west of its present location (Fig. 4a). As depicted by the tectonic reconstuctions in Figure 4, the location of the subduction trench has migrated gradually to the southeast, driven by the sinking of the slab—the Proto-Banda Sea—into the mantle through a process of 'slab rollback' [13].

The Proto-Banda Sea, Jurassic in age, once occupied a 'Banda Embayment' within the Australian continental margin not too disimilar in shape to the modern Banda Sea [5, 13]. As this old, cold, and dense oceanic lithosphere rolled back into the embayment (Fig. 3b,c), extension of the lithosphere behind the arc drove oceanic spreading, and the formation of new oceanic crust beneath the modern Banda Sea. The North Banda Basin and South Banda Basin spread open at different times behind this rolling-back arc, between 12.5–7.15 Ma [7], and 6.5–3.5 Ma [8], respectively. In addition to forming new oceanic crust, extension behind the rolling-back slab thinned and rifted apart Australian continental crust that once enclosed the northern extent of the Banda Embayment. This caused continental slivers to be standed within the Banda Sea in the form of the Banda Ridges (Fig. 2), as discovered by dredging [17]. Continental slivers may also have been transported immediately behind the rolling-back Banda Trench right up until the point of arc–continent collision between the Banda Arc and the southern Banda margin, causing Australian-affinity blocks from north of the Banda Sea to have been accreted onto a different part of the Australian continental margin in the vicinity of Timor and Babar [3, 6].

Extreme lithospheric extension driven by Banda Slab rollback (Fig. 4) also affected islands in the

IOP Conf. Series: Earth and Environmental Science **184** (2018) 012005 doi:10.1088/1755-1315/184/1/012005

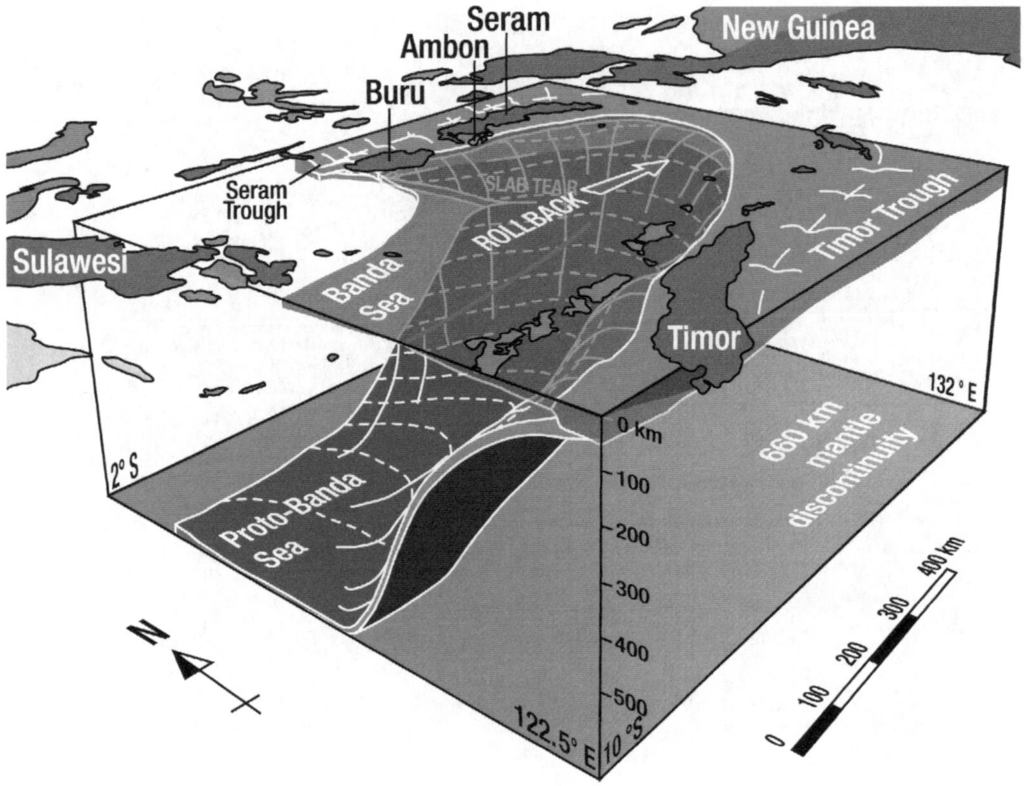

Figure 3. Present-day configuration of the Banda Sea, overlying the curved Banda Slab (the Proto-Banda Sea), modified from Pownall *et al.* (2014) [16]. Note the horizontal tear in the Banda Slab beneath Buru and western Seram.

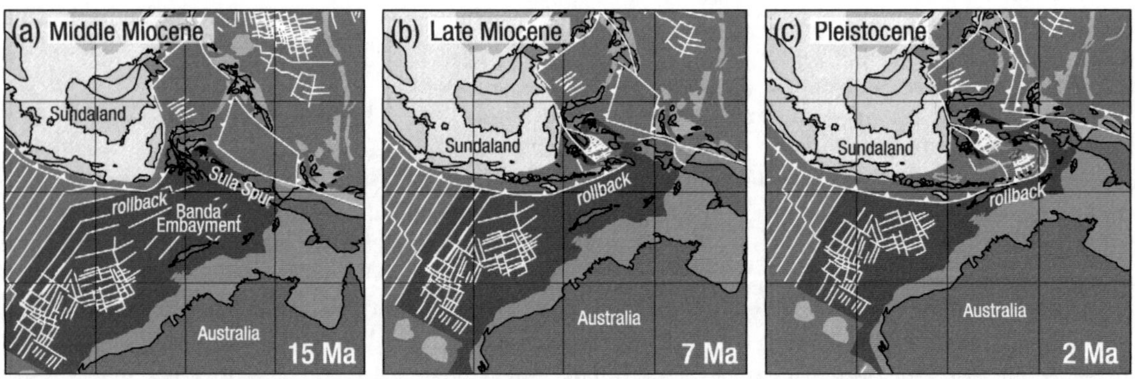

Figure 4. Tectonic reconstruction of eastern Indonesia, depicting formation of the modern Banda Sea during rollback of the Banda Arc, at (**a**) 15 Ma, (**b**) 7 Ma, and (**c**) 2 Ma (adapted from Hall, 2012 [4]). Oceanic crust is shown in dark blue (older than 120 Ma) and mid-blue (younger than 120 Ma); submarine arcs and oceanic plateaus are shown in pale blue; volcanic island arcs, ophiolites and material accreted along plate margins are shown in green.

northern Banda Arc. Hot mantle rocks were exhumed to the shallow sub-surface [11, 16, 18, 19, 20], driving crustal metamorphism of adjacent rocks under ultrahigh-temperature (UHT; > 900°C) conditions [16, 21]. Volcanic rocks comprising the island of Ambon are shown to be derived primarily from the melting of these stretched continental crustal rocks [20], and so are disctinct from the volcanic products of the rest of the arc. However, volcanoes from Banda Api to Damar also record elevated continental input due to the subduction of continental material from the Australian margin [9, 10].

Exhumation of mantle rocks and HT–UHT metamorphic rocks in the northern Banda Arc was faciliated, in part, by strike slip faults such as the Kawa Fault of Seram (Fig. 2), and the associated Kobipoto Mountains pop-up structure [18, 21, 12]. The Kawa Fault is a major structure that has been fundamental in enabling the Banda Arc to roll back eastwards with respect to the northern Banda Embayment margin. The fault also forms part of a larger structure, the Seram–Kumawa Shear Zone, which may have developed in the Jurassic when continental blocks were rifted from the Banda Embayment [12].

3. SE Asia–Australia collision and the Indonesian Throughflow

Collision of SE Asia with Australia at *c.* 23 Ma [5] closed the deepwater passageway that existed between the Indian and Pacific oceans [1, 22]. The Indonesian 'Throughflow' describes the resulting oceanic current that follows a winding and anastomosing course through the narrow and shallow Indonesian seaways from the SW Pacific to the Indian Ocean (net flow rate: 15 Sverdrups [23]). The primary route follows the Makassar Strait, Flores Sea, South Banda Sea, and Timor Trench (Fig. 1), although weaker currents also pass through the Lombok and Ombai straits. Throughflow transport is governed seasonally by El Niño–Southern Oscillation and over millennial timescales by sealevel flucuations [22, 24]. Over geological time, the configuration and strength of the Throughflow has undoubtedly been controlled by the evolving tectonic configuration, with the uplift of mountain belts on, for instance, Sulawesi restricting the flow. Due to uncertainties of reconstructing exact palaeogeographies, it is not know exactly how the passageways of the Indonesian Throughflow evolved through time [22]. However, it is thought that the passageways were especially restricted between 12 and 3 Ma, with the narrowest gaps occuring at around 10 Ma [22].

4. Isostatic controls on the Banda Sea profile

Structures within slabs beneath the Banda Sea (Fig. 5), and the mechanism by which they were subducted, have been highly influential in controling on the oceanographic profile through their isostatic response:

- Within the northern limb of the slab beneath Buru, an aseismic zone is interpreted as a tear that has caused the slab to peel away from its northern margin [13, 14, 15] (Fig. 3), shown also by tomographic models [13, 15]. Propagation of this horizontal tear may have contributed to the rapid uplift rates recorded around the northern Banda Arc [25, 26] as the upper plate rebounded.
- The southern limb of the slab features a band of intense seismicity—the Damar Zone [27]—at 100–200 km depth, which extends westwards from the Aru Trough and terminates sharply just west of Romang at the plane of intersection with the 'Gunungapi Lineament' (Fig. 2) passing from the Timor Trough to the North Banda Basin via Gunungapi Volcano. We interpret this ridge to be the surface expression of a major subvertical slab tear that delineates the actively-rupturing slab on its eastern side, and which potentially accounts for the positive increase in mean topographic elevation identified by Sandiford (2008) [27] moving west from Romang to Wetar.
- There is a shallow-dipping section of slab extending to beneath the Weber Deep [11], which likely is the down-flexed and over-thrust outermost Australian continental margin (Fig. 5). Beneath the Weber Deep is the location of the continental–ocean transition, at which point the slab steepens abpruptly. As discussed by Pownall *et al.* (2016) [11], the incredible depth of the 7.2 km Weber Deep forearc basin floor (Fig. 6) is likely supported by this shallow-dipping slab segment.

As discussed in the next section, the Weber Deep was created by the development of a low-angle detachment system during the final stages of extension behind the rolling-back Banda Slab [5, 11, 12].

5. The Weber Deep

The floor of the Weber Deep (Fig. 6, 7), beneath 7.2 km water depth, is the deepest part of Earth's oceans that does not occur in a trench. So why is this forearc basin so deep? It has been suggested that the Weber Deep formed as a flexural response to a tightening of the Banda arc's curvature [6], or in response to the thrusting of the Banda Sea over the surrounding Australian continental margin [28]. Alternatively, some authors by interpreting the feature as an extensional basin attributed east-west extension either directly to north-south shortening caused by the northward advance of Australia [29] or to eastward slab rollback [13]. The extreme depth of the basin has also been explained simply as the result of sinking of the underlying Banda slab [6, 30] without requiring rollback.

New high-resolution (15 m) bathymetry [11] has revealed in incredible detail intricate features of the eastern Banda Sea, including the Weber Deep and Aru Trough [11, 12]. Most notably, the entire Weber Deep forearc features a set of parallel striations or grooves oriented at 120–300 ± 10° (Fig. 6), which have been generated within a single low-angle detachment fault zone, the 'Banda Detachment' [11]. The detachment has a listric geometry, curving from a 12° dip adjacent to the eastern rim of the basin, becoming horizontal then slightly backrotated (by 1°) approaching the volcanic arc (Fig. 5b). The grooves' orientation and length demonstrate a southeasterly slip direction of 120–130°, along which the

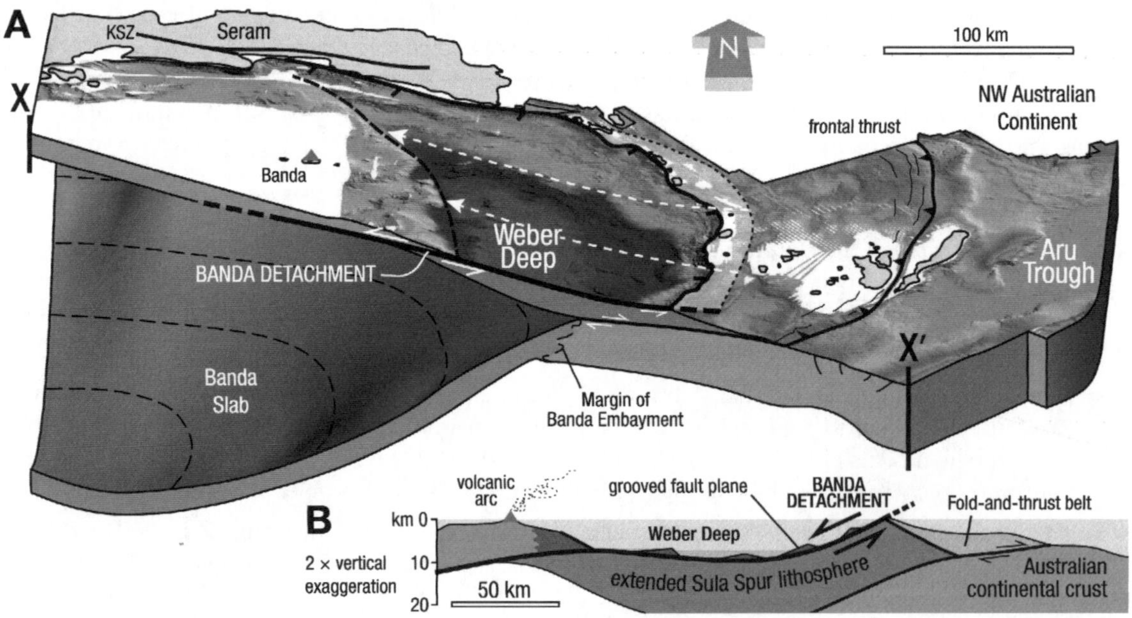

Figure 5. (**A**) Arc–continent collision in the Banda Region. Note how the continent–ocean boundary between the continental margin of the Banda Embayment (orange) and the Jurassic Proto-Banda Sea (pink) has been over-thrust by the modern Banda Sea (red and blue). Extreme lithospheric extension along the Banda Detachment has opened the Weber Deep above the Banda Slab. The location of this cut-away is shown in Fig. 6 by the line X–X′. (**B**) An enlargement of the Banda Detachment cross-section (2× vertical exaggeration) with the highly-extended Sula Spur lithosphere shown in red. Figure from Pownall *et al.* (2016) [11].

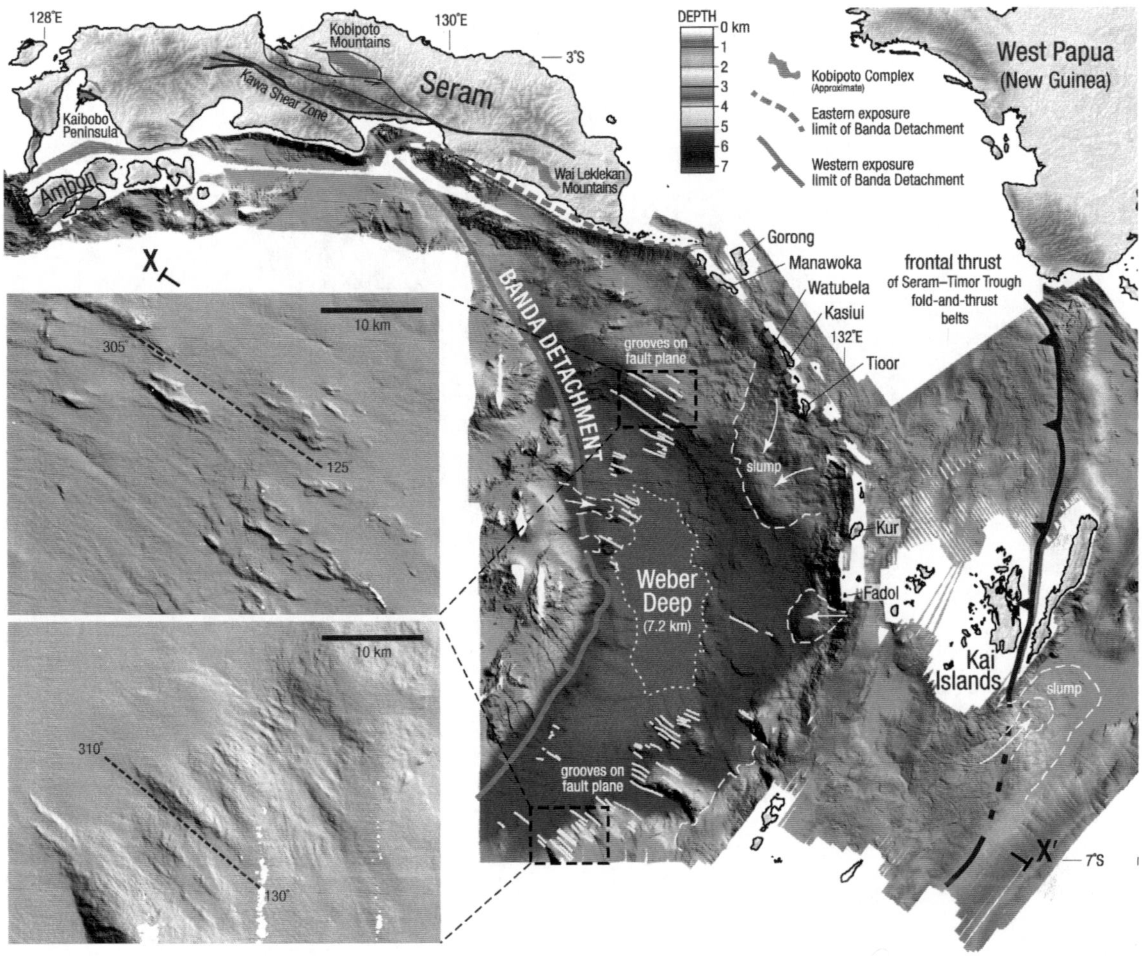

Figure 6. Bathymetric map of Weber Deep and Aru Trough (eastern Indonesia), showing the location of Banda detachment and its relationship to Kawa shear zone on Seram, after Pownall *et al.* (2016) [11]. The red areas mark approximate exposures of exhumed upper-mantle–lower-crustal (Kobipoto Complex) rocks. Multibeam data (15 m resolution) courtesy of TGS (www.tgs.com) and GeoData Ventures (Singapore). The enlargements, bottom–left, show the parallel lineations present on the Banda Detachment scarp. Note how the grooved fault scarps exposed in the Weber Deep are parallel also to the Kawa Shear Zone on Seram.

450 km-long detachment must have slipped $>$ 120 km during a massive extensional phase in the arc's evolution. The Banda Detachment is the largest identified normal fault system exposed anywhere in the world's oceans [11].

Ultramafic rocks and high-temperature metamorphic rocks (diatexites, gneisses, high-grade amphibolites) are exposed all around the northeastern rim of the Weber Deep, from the Wai Leklekan Mountains of eastern Seram, to the small islands of Kur and Fadol (Fig. 8) in the east of the arc. Offset along a major normal fault is the only plausible way of explaining how these lower-crustal–upper-mantle rocks are exposed adjecent to a 7 km-deep basin [11]. Furthermore, the Banda Detachment fault scarp—dipping into the Weber Deep at a consistent 12°—has been observed on Fadol (Fig. 8) and in eastern Seram [11]. There is also a connection between the Banda Detachment and the Kawa Fault Zone, described previously. Both the Kawa Fault and the grooves on the Banda Detachment fault scarps are

IOP Conf. Series: Earth and Environmental Science **184** (2018) 012005 doi:10.1088/1755-1315/184/1/012005

Figure 7. Perspective view of the Weber Deep (large vertical exaggeration; looking north) showing the large submarine landslides and submerged pinnacle reefs.

Figure 8. Sailing south to the island of Fadol in the eastern Banda Arc. The Banda Detachment bounds the gently-dipping western side of the island (right side in the photo). On Fadol, high-temperature metamorphic rocks and ultramafic rocks are exposed, capped by uplifted carbonate terraces.

parallel, thereby demonstrating the two structures are in some way coupled; and so likely acted together to facilitate southeastward slab rollback plus extreme forearc extension.

The oceanographical profile of the Weber Deep has been also influenced by submerged pinacle reef structures (alternatively mud volcanoes?; Fig. 7), and by large submarine debris flows that blanket much of the eastern rise (Fig. 6,7). These flows, some continuous over 100 km, demonstrate the mass transport of unstable material from the shallow shelf into the abyss. They also provide evidence that the Banda Detachment is either active, or only recently ceased.

From a geohazards perspective, these mass debris flows may pose a greater tsunami risk than earthquakes produced by the vast Banda Detachment. As the Banda Detachment is now exposed at the seabed, it can no longer generate earthquakes other than beneath the volcanic arc, which comprises its hanging wall (Fig. 5b). Nevertheless, frequent low-magnitude (< 5) shallow (< 60 km) earthquakes recorded in the eastern Banda Sea suggest that steeper-angle faults beneath the Banda Detachment, and dominantly strike-slip faults at the Weber Deep's northern and southern extents, faciliate continued extension.

References

[1] Meyers G, Bailey R J and Worby A P 1995 *Deep-Sea Research Part I-Oceanographic Research Papers* **42** 1163–1174
[2] Gordon A L and Kamenkovich V M 2010 *Dynamics of Atmospheres and Oceans* **50** 113–114
[3] Hall R, Cottam M A and Wilson M E J 2011 *Geological Society, London, Special Publications* **355** 1–6
[4] Hall R 2012 *Tectonophysics* **570-571** 1–41
[5] Hall R 2011 *Geological Society, London, Special Publications* **355** 75–109
[6] Bowin C, Purdy G M, Johnston C, Shor G, Lawyer L, Hartono H and Jezek P 1980 *AAPG Bulletin* **64** 868–915
[7] Hinschberger F, Malod J A, Réhault J P, Dyment J, Honthaas C, Villeneuve M and Burhanuddin S 2000 *Comptes Rendus de l'Académie des Sciences - Series IIA - Earth and Planetary Science* **331** 507–514
[8] Hinschberger F, Malod J A, Dyment J, Honthaas C, Rehault J P and Burhanuddin S 2001 *Tectonophysics* **333** 47–59
[9] Vroon P Z, van Bergen M J, White W M and Varekamp J C 1993 *Journal of Geophysical Research* **98** 22349–22366
[10] Nebel O, Vroon P Z, van Westrenen W, Iizuka T and Davies G R 2011 *Earth and Planetary Science Letters* **303** 240–250
[11] Pownall J M, Hall R and Lister G S 2016 *Geology* **44** 947–950
[12] Hall R, Patria A, Adhitama R, Pownall J M and White L T 2017 *Proceedings of the Indonesian Petroleum Association* **41** IPA17–91–G
[13] Spakman W and Hall R 2010 *Nature Geoscience* **3** 562–566
[14] Pownall J M, Hall R and Watkinson I M 2013 *Solid Earth* **4** 277–314
[15] Hall R and Spakman W 2015 *Tectonophysics* **658** 14–45
[16] Pownall J M, Hall R, Armstrong R A and Forster M A 2014 *Geology* **42** 279–282
[17] Silver E A, Gill J B, Schwartz D, Prasetyo H and Duncan R A 1985 *Geology* **13**
[18] Pownall J M and Hall R 2014 *Proceedings of the Indonesian Petroleum Association* **38**
[19] Pownall J M, Forster M A, Hall R and Watkinson I M 2017 *Gondwana Research* **44** 35–53
[20] Pownall J M, Hall R and Armstrong R A 2017 *Gondwana Research* **52**
[21] Pownall J M 2015 *Journal of Metamorphic Geology* **33** 909–935
[22] Kuhnt W, Holbourn A, Hall R, Zuvela M and Käse R 2004 *Neogene History of the Indonesian Throughflow* (*Geophysical Monograph Series* vol 149) (Washington, D. C.: American Geophysical Union)
[23] Gordon A L, Sprintall J, Van Aken H M, Susanto D, Wijffels S, Molcard R, Ffield A, Pranowo W and Wirasantosa S 2010 *Dynamics of Atmospheres and Oceans* **50** 115–128
[24] Tillinger D 2011 *Geological Society, London, Special Publications* **355** 267–281
[25] Fortuin A R, de Smet M E M, Sumosusastro P A, Van Marle L J and Troelstra S R 1988 *Geologie en Mijnbouw* **67** 91–105
[26] de Smet M E M, Fortuin A R, Tjokrosapoetro S and Vanhinte J 1989 *Netherlands Journal of Sea Research* **24** 263–275
[27] Sandiford M 2008 *Geophysical Journal International* **174** 659–671
[28] Hamilton W 1979 *USGS Professional Paper* **1078** 345
[29] Charlton T R, Kaye S J, Samodra H and Sardjono 1991 *Marine and Petroleum Geology* **8** 62–69
[30] McCaffrey R 1988 *Journal of Geophysical Research: Solid Earth* **93** 15163–15182

A hard way towards sustainable management of Banda Sea ecosystems and their energy resources

S Susilohadi[1]

[1]Research and Development Centre of Marine Geology
Jalan Dr. Junjunan 236 Bandung-40174, Indonesia

s.susilohadi@mgi.esdm.go.id

Abstract. The Banda Sea region which includes the Moluccas and Nusa Tenggara consists of 3,478 small islands separated from each other by deep seas. This region can be regarded as the most complex energy supply chain in Indonesia due to geographic condition and lacking of energy infrastructure. The newly issued Indonesian general plan of energy aims to establish a strategy for planning the region's energy mix until 2050. By then, national primary energy should not be dominantly supported by oil and gas, and renewable energy should contribute 31%. In addition, government has committed to attain near 100% electricity ratio by 2020. Recent hydrocarbon discoveries and ocean energy mapping indicate that the Banda Sea region is an attractive area where the country's primary energy needs may partly be fulfilled. However, the region's sustainability in energy production still has to be established with greater efforts. Hydrocarbon search has been discouraged by the geologically risky nature of the region. The implementation of renewable energy, particularly ocean energy, is hampered by technology immaturity and constraining operational rules and policies.

Keywords: Banda Sea, energy consumption, hydrocarbon resources, ocean energy, energy policy

1. Introduction

The Banda Sea region which includes the Moluccas and Nusa Tenggara is unique, consisting of 3,478 small islands separated from each other by deep seas (Table 1). This region that has population of more than 11,755,632 people [1], may be regarded as having the most complex energy supply chains in Indonesia due to such a geographic condition, in addition to the lacking of energy infrastructure including network, storage facilities and tanker fleets to ensure their continued reliability and efficiency.

This year, Indonesian Government has issued the new general plan of national energy (RUEN) [2] which is intended as a guide for ministries, governmental institutions and local governments to plan and establish their strategy and planning on energy and to the community and private sectors on how they can participate in the energy developments. For the Banda Sea region where electrification ratio (population's access to electricity) is below 85%, this RUEN is a kind of government commitment since 100% ratio has to be reached within the next 3 years. In line with the national policy on energy [3], the uses of new and renewable energy sourced locally are made necessary by the archipelagic nature of Indonesia.

The Indonesian hydrocarbon production is currently in a declining state since 1995, as no new reserve has been found in at least last 10 years [4]. Meanwhile, eastern Indonesia is considered to be

Content from this work may be used under the terms of the Creative Commons Attribution 3.0 licence. Any further distribution of this work must maintain attribution to the author(s) and the title of the work, journal citation and DOI.
Published under licence by IOP Publishing Ltd

underexplored for hydrocarbons [5]. On the other mode of energy, the National Energy Council [6] indicates that Indonesian new and renewable ocean energy opportunities are abundant. Most of them are located in the Banda Sea region [7, 8]. These circumstances have led to an expectation that future energy reserves may rest in the eastern Indonesia.

This paper reviews the current state of energy resources in the Banda Sea region and discusses the energy related policies which could impact regional development. Ocean energy is highlighted since potentially would be part of the sustainable ecosystem of the region.

2. Statistics

Statistics provided by the upstream regulator agency of oil and gas [1] indicate that fuel consumption in the provinces around Banda Sea are very low, less than 1% of the national consumption (table 1) which reached more than 71.5 billion liters in 2015. Electrical consumption of the Banda Sea region also shows similar figures, with more than 90% used by residential consumers (table 2) [9]. Currently, electricity is generated by diesel or coal-fired steam power sourced outside the region.

This low consumption of energy may be due to the fact that the regional economic support to the national GDP is low, about 19%, which is mainly come from by mining industries [10]. Geographic condition and lack of infrastructure may have lead to the low non-mining industrial development in eastern Indonesia, as indicated by Competitiveness Industrial Performance (CIP) below 0.1% [1]. However, economic assessment by Bank Indonesia indicates that some provinces in eastern Indonesia are ready to develop such industries, such as cacao, palm oil and fishing industries, although slowing down of economic growth is also recorded [10].

Table 1. Demography and fuel consumption statistics in the Banda Sea region.

No.	Provinces	Islands	Population (2010)	Economic growth 2016	Fuel consumption 2015 (Liter)	% of national consumption
1.	North Moluccas	1,474	1,038,087	5.8%	270,819,144	0.38%
2.	Moluccas	1,422	1,533,506	5.8%	415,307,408	0.58%
3.	West Nusa Tenggara	864	4,500,212	5.8%	76,6201,764	1.07%
4.	East Nusa Tenggara	1,192	4,683,827	5.2%	485,87,0450	0.68%

Table 2. Electricity consumption statistics (as of 2015) in the Banda Sea region.

No.	Provinces	Electrification ratio	Installed Capacity (MW)	Diesel (MW)	Coal (MW)	Residential customer	Industrial & Commercial
1.	North Moluccas	94.46%	58.00	51.00	7.00	91.78	4.81%
2.	Moluccas	84.80%	159.01	158.12		91.39%	5.77%
3.	West Nusa Tenggara	72.77%	166.52	110.29	50.00	94.31%	3.23%
4.	East Nusa Tenggara	58.64%	216.79	153.82	47.00	92.29%	5.04%

3. Energy resources in the Banda Sea region

3.1. Hydrocarbons

The geology of eastern Indonesia results from a complex interaction between Indo-Australian, Western Pacific, Philippine Sea and Eurasian plates. The relative NNE movement of the Indo-Australian plate and westward movement of the Philippine Sea and Western Pacific plates since Late Oligocene have formed the horseshoe-shaped of Banda arc which extends from Flores to Buru, including Timor and Seram [11]. In contrast to western Indonesia, which is largely underlain by continental crust, eastern Indonesia is dominated by arc and ophiolitic crust, and several young ocean basins.

Petroleum systems are well developed in the Banda arc area. There are more than 25 sedimentary basins and 50 active petroleum company working areas in eastern Indonesia (figure 1). Seram, Bula and Salawati are well known oil producing areas that have been operating since the Dutch colonialization. The newer discoveries are in Bintuni basin (mid 1990s) and Masela Block (early 2000s), which reserve 14 and 18 TCF of gas, respectively [5]. Surprisingly, compared to the western Indonesian hydrocarbon basins which are mostly of Tertiary age, in both Bintuni and Masela gas was found within pre-Tertiary rocks and may be regarded as a new insight in hydrocarbon searching in eastern Indonesia [12]. Similarly, gas was also found along the margin of Sahul Shelf in the Australian territory (southeast off Timor Leste), such as in Bayu-Undan and Elang-Kakatua Fields.

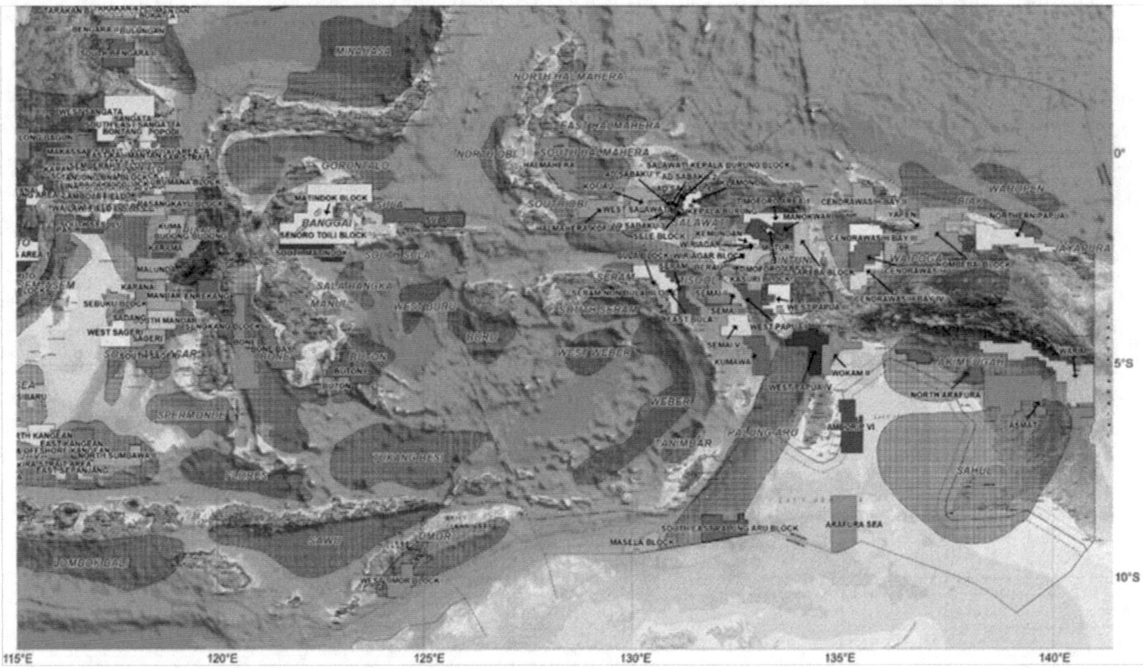

Figure 1. Sedimentary basins (shaded grey) and hydrocarbon working areas (colored boxes) in eastern Indonesia, drawn based on the data from the Data and Information Center, Indonesian Ministry of Energy and Mineral Resources (Pusdatin ESDM).

Despite the success, there are some unsuccessful exploration wells in the eastern Indonesia. Up to 2013, twelve contractors drilled 25 exploration well valued US$ 1.9 billion (table 3). All of them were dry, found uneconomic prospects or suffered technical problems [13]. These have indicated that eastern Indonesia is indeed geologically complex and cannot be studied partially based on limited data. High quality and denser spacing of seismic data are absolutely required in order to minimize risk and cost of the following exploration stages. Satyana *et al* [14] underlined that the earlier oil production successes in late 1970s and mid 1990s were due to massive exploration of almost ten years earlier.

3.2. Ocean energy

Ocean energy refers to renewable energy obtained from the sea, either as mechanical energy from the tide and waves or as thermal energy from the sun. Tidal energy arises from tidal movements, utilising the vertical changes in sea levels or the horizontal movement of the seas and currents to generate electricity. Wave energy uses the energy of ocean waves or swells to generate electricity. Ocean thermal energy conversion (OTEC) uses the difference in temperature between the warmer surface or shallow waters and the cooler deeper waters to generate electricity. Practical temperature difference is in the order of 20 °C [15], which is usually found on areas with water depth of more than 1000 m.

Table 3. Unsuccessful hydrocarbon exploration drilling in eastern Indonesia by various operators [13].

No	Operator (KKKS)	Working area	Well name	Spending (million US$)	Well status
1	Exxon Mobil	Surumana	Rangkong-1	123	Dry
			Kris-1	45	Biogenic Gas Uneconomic
2	Exxon Mobil	Mandar	Sultan-1	110	Dry
			Kriss Well-1 ST	24	Dry
			Gatotkaca-1 ST	98	Dry
3	Statoil	Karama	Anoman-1	43	Dry
			Antasena-1	33	Dry
		Kuma	Kaluku-1	150	Waxy Oil (MDT)
4	Conoco Phillips	Amborip VI	Aru-1	58	Dry
		Arafura Sea	Mutiara Putih-1	103	Dry
5	Talisman	Sageri	Lempuk-IX	84	Dry
			Bravo Well	103	Dry
6	Marathon	Pasang Kayu	Romeo Well	23	Dry
			Romeo B-1	25	Technical Problem
			Romeo C-1	58	Technical Problem
7	Tately	Budong-Budong	KD-1	34	Technical Problem
			LG-1	17	Uneconomic Well
8	Japex	Buton	Benteng-1	31	Dry
9	CNOOC	SE Palung Aru	Sindoro-1	50	Dry
10	Hess	Semai IV	Andalan-1	164	Dry
			Andalan-2	59	Dry
11	Niko Resources	Kofiau	Ajek-1	37	Sub Commercial Gas Discovery
		W. Papua IV	Cikar-1	87	Temporarily Suspended
		N. Makassar Strait	Pananda-1	90	Drilling
12	Murphy Oil	Semai II	Lengkuas-1	215	Dry
			Total	1900	

The Indonesian Ministry of Energy and Mineral Resources (ESDM) has mapped and carried out assessment of Indonesian marine energy potentials. The results are indicated in the general planning of national energy [2]. More than 17 GW of energy may potentially be harvested from Indonesian ocean. Mapping by the Research and Development Centre of Marine Geology since 2006 found even larger results [16]. Tidal current measured on narrow straits between islands in Nusa Tenggara and northern West Papua may reserve 17.9 GW of energy, while waves modelled along coastline southwest of Java and Sumatra, and south of Papua result in 1.9 GW of energy (figure 2) [17].

Ocean thermal means temperature different at about 20°C between surficial and lower ocean layers and usually occurs in an ocean which has the water depth of more than 1000 m. Many islands in Indonesia are directly adjacent to the water depth of 1000 m or more, particularly in eastern Indonesia. By taking into account the length of the coastline of such islands, the Agency estimates that more than 41 GW of energy can be harvested from this temperature different (figure 3) [18].

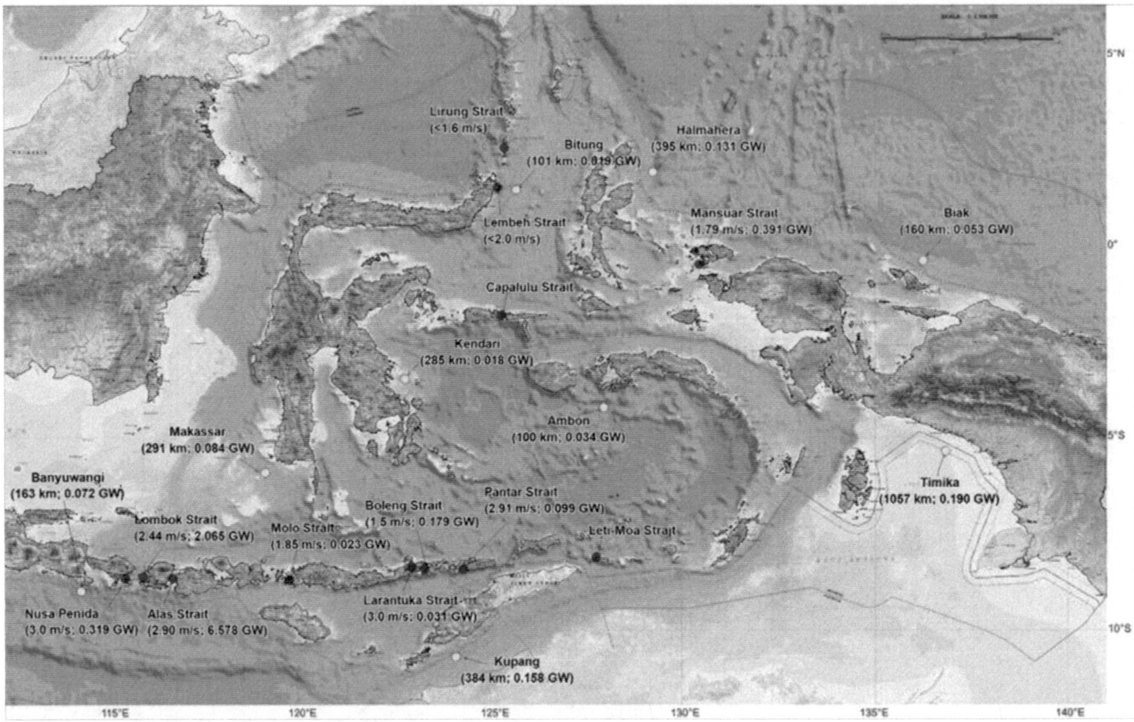

Figure 2. Map showing locations of field tidal current mapping (blue dots) and wave modelling (yellow dots) in Nusa Tenggara, Papua and Sulawesi, and their corresponding potential energy [16, 17].

4. Discussion

4.1. Hydrocarbon resources

Indonesian oil reserves are slowly declining by 1.2% and no longer sustainable as about 44% of national fuel demand is imported. Most of the known reserves are located in traditional places, such as in Sumatra, Kalimantan, and Java. Papua, Maluku, and Sulawesi may hold only 2.33% of national reserve, even though, gas is proven to be prolific. In line with economic and population growth, national energy demand is steadily growing, so the search for new hydrocarbon prospect is inevitable. Eastern Indonesia, despite being a geologically complex region, has been seen as a new frontier.

Most of eastern Indonesia has been considered to be underexplored due to lack of geological and geophysical data, its deep water nature and poor infrastructure. New hydrocarbon findings, such as in the Masela Block, suggest that the way they were generated and accumulated are different than in western Indonesia. Pre-Tertiary rocks should become exploration targets, although they have usually been involved in a complex geological history. Comprehensive geological studies are needed and to understand such a geologically complex region, good quality and denser geological and geophysical data are absolutely required. Multi-client seismic data acquired by private companies are already available; however, only few studies have been carried out based on these data as they can only be accessed commercially. Meanwhile, there is no government institution capable of providing such good quality data. It is suggested that the Indonesian government should provide stronger support to geological research institutes and universities to study eastern Indonesian geology comprehensively by providing better data acquisition facilities and access on any available data.

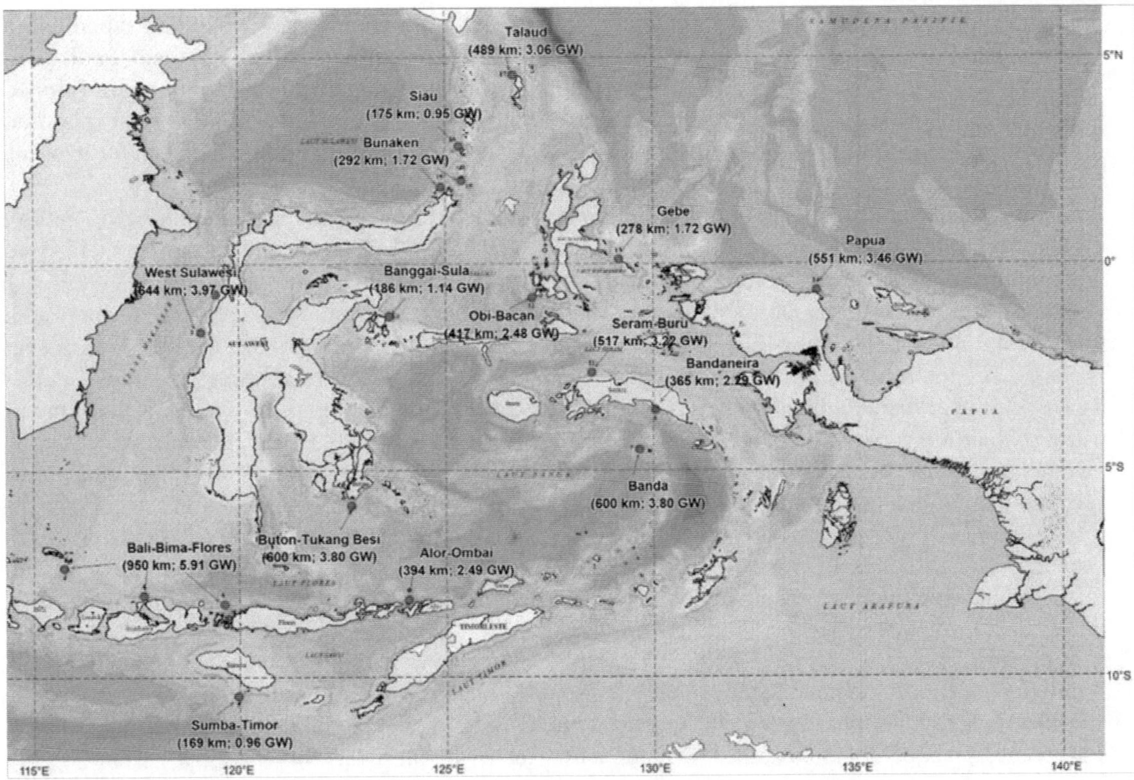

Figure 3. Map of ocean thermal energy potential in eastern Indonesia. Red dots represent location where measurement of coastline length adjacent to water depth >1000 m (yellow lines) was taken. Text accompanied dots indicates length of coastline and calculated practical energy potential [18].

4.2. Marine energy resources

In the new general plan of national energy [2] Indonesia has targeted that renewable energy has to share at least 23% (or 45.2 GW) of total primary energy supply by 2025, and at least 31% by 2050. Renewable energy such as ocean energy should contribute 3.1 GW along with bioenergy. Recently, the ESDM issued a decree [19] where operational steps of electricity procurement will be made as close as possible to those energy mix.

Official data has indicated that opportunities for ocean energy production are potentially high in the Banda Sea region. In 2015, the ESDM has indicated that Indonesia would encourage the use of energy potential from the sea as part of the government's marine development policy [20]. The ministry currently has two pilot projects underway - one in Nusa Penida and another in Nusa Tenggara. In line with that, the state oil company (Pertamina) has committed to develop 3 MW of ocean energy by 2019. Potential energy assessment of the Alas Strait, Lombok Strait and Badung Strait are currently on going, as the state electricity company (PLN) intends to develop ocean energy from a capacity of 12 to 140 MW [21].

Despite these strong intentions, the readiness of marine energy to support the national energy mix will unlikely to go as fast as expected. At least three major issues which may discourage the development are identified:

- Conversion technology may be site-specific due to hydro-oceanographic nature of eastern Indonesian waters and hence the technology that should be applied accordingly. Speed of tidal currents in straits of Indonesia is rarely reach 3.0 m s^{-1}, and strongly influenced by semi-diurnal nature of tides [16]. In addition, the nature of Indonesian narrow straits is very different to one another. Straits in Bali and Nusa Tenggara, for example, are characterized by steep slope and a rough sea floor, since their formations were influenced by volcanic deposits and geological structures. Most of the current technology to convert marine energy into electricity are developed and tested in western European and northern America where ocean current is much stronger influenced by diurnal tides. Such technology if to be applied in Indonesian water may need some technological adjustments.

- Under ministerial decree [19], PLN will consider to buy electricity converted from ocean energy if the technology has been proven commercially reliable for 5 years. However, so far there is no single prototype has been running or even proved to be reliable in Indonesian water.
- The new ministerial regulation [22] sets the policy on tariff capping of renewable energy. It regulates the framework on how PLN procures electricity from renewable energy-based power plants. All renewable energy tariffs will be capped at 85% of the local generation cost, if this price is higher than the national average generation cost. If the local generation cost is the same or lower than the national average, then the limit will be 100% of the local generation cost. The regulation asks that proven conversion technology has to compete with other modes of conversion which may have already been mature technologically, such as hydro and hydrocarbon-burning electricity generators. Currently, ocean energy conversion technology is far from being mature and, Indonesian research institutes required to develop such technology have received severe research budget cuts.

5. Conclusion

The Indonesian energy demand steadily increas proportional to economic growth. The steady decline of hydrocarbon production has led the government to issue a new general plan of national energy which involves reducing the country's dependence on hydrocarbons. New and renewable energy sources are now becoming the national target for exploration and exploitation. Nevertheless, Indonesian primary energy will still lean on oil and gas as much as 44% by 2050.

The Banda Sea region comprises provinces that are expected to be the next massive exploration target for hydrocarbons. The region is also rich in renewable energy, particularly ocean energy. Unfortunately, the region's sustainability in energy may still have to be established with greater efforts. Hydrocarbon exploration has been discouraged by the geologically risky nature and lack of supporting infrastructures of the region. On the other hand, practical implementation of renewable energy, such as ocean energy, is hampered by technology immaturity and constraining operational rules and policies.

Acknowledgements

The author wish to thank Dr. Augy Sahilatua and Dr. Yosmina Tapilatu for the invitation and support to be involved in the International Symposium on Banda Sea Ecosystem 2017 on which this manuscript has been intended to. The author also would like to appreciate the reviewers for the corrections, comments and suggestions on the manuscript.

References

[1] Badan Pusat Statistik 2015 *Statistik Indonesia: Statistical yearbook of Indonesia 2015* (Jakarta: Badan Pusat Statistik) pp 710

[2] Kementerian Hukum dan Hak Asasi Manusia 2017 *Peraturan Presiden Republik Indonesia nomor 22 tahun 2017 tentang Rencana Umum Energi Nasional* (Jakarta: Sekretariat Kabinet RI)

[3] Kementerian Hukum dan Hak Asasi Manusia 2014 *Peraturan Pemerintah Republik Indonesia nomor 79 tahun 2014 tentang Kebijakan Energi Nasional* (Jakarta: Kementerian Sekretariat Negara RI)

[4] Direktorat Jenderal Minyak dan Gas Bumi 2015 *Rencana strategis 2015-2019* (Jakarta: Kementerian Energi dan Sumber Daya Mineral, Direktorat Jenderal Minyak dan Gas Bumi)

pp 74
[5] Satyana A H 2013 Oil and gas in eastern Indonesia: production, proven plays, prospectivity, and risks *AAPG-UGM Student Chapter, Guest Lecture, Yogyakarta, 10 March 2013*
[6] Dewan Energi Nasional 2016 *Indonesia energy outlook 2016* (Jakarta: Sekretariat Jenderal Dewan Energi Nasional) pp 83
[7] Badan Pengkajian Dan Penerapan Teknologi 2014 *Outlook energi Indonesia 2014 pengembangan energi untuk mendukung program substitusi BBM* ed A Sugiyono et al (Jakarta: Pusat Teknologi Pengembangan Sumberdaya Energi Center For Energy Resources Development Technology, Agency For The Assessment And Application Of Technology) pp 129
[8] Badan Pengkajian Dan Penerapan Teknologi 2016 *Outlook energi Indonesia 2014 pengembangan energi untuk mendukung industri hijau* ed A Sugiyono et al (Jakarta: Pusat Teknologi Sumber Daya Energi Dan Industri Kimia, Badan Pengkajian Dan Penerapan Teknologi) pp 119
[9] Direktorat Jenderal Ketenagalistrikan 2015 *Statistik ketenagalistrikan 2015* (Jakarta: Direktorat Jenderal Ketenagalistrikan, Kementarian Energi dan Sumber Daya Mineral) pp. 41
[10] Bank Indonesia 2016 Kajian ekonomi dan keuangan regional *Laporan Nusantara* 11 4 pp 91
[11] Hall R 2009 *Encyclopedia of islands: Indonesia, geology* ed Gillespie R. and Clague D (Berkeley: University of California Press) p. 454-460
[12] Robertson J D 2002 Tangguh-the first major pre-Tertiary discovery in Indonesia *AAPG Annual Meeting, Houston, March 10-13, 2002*
[13] Sunardi L 2013 Pemerintah putus kontrak Marathon Oil di Blok Pasang Kayu *Bisnis Indonesia, 30 Juni 2013*
[14] Satyana A H, Armandita C and Paju J A 2012 Acceleration in regional exploration of Indonesia: Requirement for survival *Proceedings, Indonesian Petroleum Association, Thirty-Sixth Annual Convention and Exhibition, May 2012*
[15] Nihous G C 2005 An order-of-magnitude estimate of ocean thermal energy conversion resources *Transactions of the ASME, vol. 127, December 2005* p. 328-333
[16] Pusat Penelitian dan Pengembangan Geologi Kelautan 2014 *Peta potensi energi arus laut Indonesia* (Bandung: Pusat Penelitian dan Pengembangan Geologi Kelautan, Badan Penelitian dan Pengembangan Energi dan Sumber Daya Mineral)
[17] Pusat Penelitian dan Pengembangan Geologi Kelautan 2014 *Peta potensi energi gelombang laut Indonesia* (Bandung: Pusat Penelitian dan Pengembangan Geologi Kelautan, Badan Penelitian dan Pengembangan Energi dan Sumber Daya Mineral)
[18] Pusat Penelitian dan Pengembangan Geologi Kelautan 2014 *Peta potensi energi panas laut Indonesia* (Bandung: Pusat Penelitian dan Pengembangan Geologi Kelautan, Badan Penelitian dan Pengembangan Energi dan Sumber Daya Mineral)
[19] Kementerian Energi dan Sumber Daya Mineral 2017 *Keputusan Menteri Energi dan Sumber Daya Mineral Republik Indonesia no. 1415 K/20/MEM/2017 tentang Pengesahan rencana usaha penyediaan tenaga listrik PT Perusahaan Listrik Negara (persero) tahun 2017 s.d. 2026* (Jakarta: Kementerian Energi dan Sumber Daya Mineral)
[20] Kementerian Energi dan Sumber Daya Mineral 2015 *Renstra KESDM 2015-2019* (Jakarta: Kementerian Energi dan Sumber Daya Mineral) pp. 275
[21] PwC 2016 Power in Indonesia, investment and taxation guide: November 2016 - 4th edition (Jakarta: PwC Indonesia) pp. 180
[22] Kementerian Energi dan Sumber Daya Mineral 2017 *Peraturan Menteri Energi dan Sumber Daya Mineral Republik Indonesia no. 12/2017 tentang pemanfaatan sumber energi terbarukan untuk penyediaan tenaga listrik* (Jakarta: Kementerian Energi dan Sumber Daya Mineral)

Current Status and Future Prospect of Marine Pollution Research in the Banda Sea

C Y Manullang[1]

[1]Centre for Deep-Sea Research, Indonesian Institute of Sciences (PPLD LIPI), Jl. Y. Syaranamual, Guru-Guru Poka Ambon 97233, Indonesia.

corry.yanti.manullang@lipi.go.id

Abstract. Anthropogenic pressure can cause serious problem in the marine environment, such as marine pollution. In this paper the existing scientific literatures related to marine pollution study that had been conducted at the Banda Sea were reviewed. There were in total 22 publications found that specifically addressed the marine pollution at the Banda Sea. The majority of case studies were conducted in shallow coastal waters. Few data discussed about deep-sea waters. We highlighted knowledge gaps regarding the information on data from Banda Islands coastal area. The final goal of this paper is to recommend the future studies that need to be conducted in the Banda Sea waters.

Keywords: *marine pollution, Banda Sea, deep sea*

1. Introduction

The Banda Sea is the deepest sea in Indonesia that is connected to the Pacific Ocean and surrounded by hundreds of islands (figure 1). The deepest part is the Weber Trench which is 7 440 meters deep [1]. The Banda Sea is also located at the heart of the Coral Triangle that makes this area the highest marine biodiversity spot in Indonesia. In 2014, Indonesian Ministry of Maritime Affairs and Fisheries estimated that the total capture fisheries production of Maluku waters was at 538 thousand tons [2]. Due to this large potential, the government of the Republic of Indonesia has declared the Moluccas Province to be a "National Fish Barn" since 2014.

However, like in the other part of the world, the Banda Sea could be exposed to anthropogenic activities. One of the anthropogenic stresses that can be a serious problem in the marine environment is marine pollution. The marine pollution occurs when any substances or energy enter the marine environment and bring out the harmful effects upon ecosystems and human health [3].

The marine pollution study provides information about the disturbances of the oceanic environment and its biota, as well as its effects on the environment and human safety. In this paper, we reviewed the marine pollution study conducted in and around the Banda Sea since the first publication in 1987 until 2017. We identified any research gap and propose the next marine pollution research that needs to be done in the Banda Sea.

2. Marine Pollution Investigation in the Banda Sea, 1983-2017

Our understanding of the present status of marine pollution in the Banda Sea is surprisingly limited. Most of the studies that have been conducted were concentrated in Ambon Island that is

Content from this work may be used under the terms of the Creative Commons Attribution 3.0 licence. Any further distribution of this work must maintain attribution to the author(s) and the title of the work, journal citation and DOI.
Published under licence by IOP Publishing Ltd

predominantly shallow with <200 m in depth. There is only one published data about marine pollution in Banda waters. Even though the data was limited, a number of studies on marine pollution have been conducted in Ambon Island. At least, from 1983 to 2017 (table 1), there were 22 publications in the Banda Sea consisting of 17 studies about heavy metals (9 peer-reviewed articles, 7 undergraduate reports, and 1 dissertation), 3 studies on domestic waste (2 journals, one newspaper article) and 2 studies on hydrocarbon (1 peer-reviewed article and 1 undergraduate report).

2.1. Heavy Metals
2.1.1 Sea Water
The Ambon Bay consists of an inner and outer area that is connected by a narrow and shallow threshold with an average depth of about 12 m (figure 1). The outer bay area is less populated than inner area. The inner bay has an average depth of 30 m with about 6 km^2 of area. The outer bay opens into Banda Sea, and has an area of about 100 km^2 with an average depth over 100 m (figure 1).

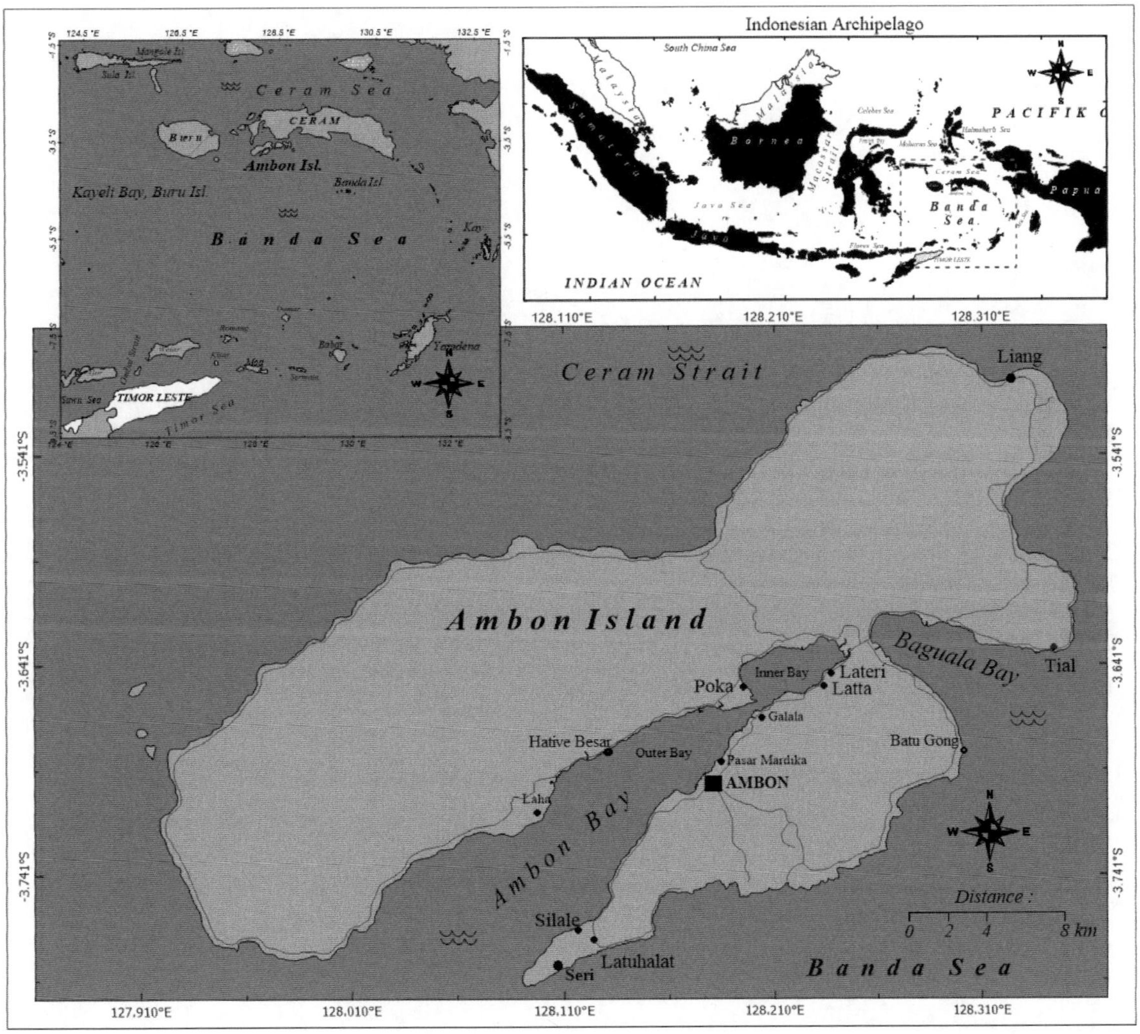

Figure 1. Map of the Banda Sea.

Heavy metal studies on sea water have been conducted in the Banda Sea since early 1980s. Two investigations in 1983 and 1987 on sea water found high concentrations of Cu, Zn, Pb and Cd in Ambon Bay [4, 5] (table 1). The range of Cu, Zn and Cd in Ambon Bay has exceeded the limits of the national standard on marine water quality (Decree of the Indonesian Minister of environment no. 51, 2004) [6]. After more than two decades since the first publication about Ambon Bay, the total concentrations of Cu, Zn, Pb, Cd and Hg in seawater from this bay were reported to be below the upper limit [7] than those observed in 1987 irrespective of population density. The decreased trend 1983 to 2009 might be caused by many factors, such as: sampling time (seasons), different procedure in metal determinations and bias in methodological approach. Unfortunately, these studies were difficult to compare due to the lack of methodological details in the published paper and different sampling time for each study.

Table 1. Heavy metals studies in the Banda Sea, 1983-2017 (in $mg.kg^{-1}$, dry weight for sediment and biota, in $mg.l^{-1}$ for sea water). Bolded numbers indicate values exceeding the guideline's upper threshold limit for the specific heavy metal. AIB: Ambon Inner Bay, AOB: Ambon Outer Bay.

No	Location	Cr	Fe	Cu	Zn	Pb	Cd	Hg	Ref
Sea water									
1	Ambon Bay	-	-	**0.008-0.050**	0.02-**0.06**	0.035-0.075	<0.001-**0.05**	<0.001	[4]
2	Ambon Bay	-	-	0.005-**0.058**	0.009-**0.071**	0.004-**0.09**	**0.005-0.065**	-	[5]
3	Ambon Bay	-	-	<0.001	<0.0078-0.0079	<0.001-0.005	<0.001	<0.001	[7]
4	Ambon Island	-	-	-	-	-	**0.01-0.03**	-	[8]
5	AIB	-	-	-	-	**0.05-0.1**	-	-	[9]
Marine Sediment									
6	Ambon Bay	-	-	19.9-27.4	62.1-95.6	10.4-23.1	<1	0.0396-0.1225	[4]
7	Ambon Bay	-	-	0.05-0.216	0.168-10.45	0.17-0.683	0.129-**2.161**	0.05-**0.16**	[11]
8	Ambon Bay	-	-	-	-	9.473-24.901	0.163-0.444	**1.991-4.521**	[12]
9	Ambon Bay	-	-	-	1.258-48.8	1.397-15.280	<0.0013-0.512	<0.0013	[7]
10	Ambon Island	-	-	-	-	-	0.17-0.32	-	[8]
11	AIB	-	-	-	-	5-20	-	-	[9]
12	Ambon Bay	-	27,598–51,716	14.4-24.5	51.3-**163**	14.4-24.5	0.1 – 0.66	0.04–**0.44**	[13]
13	Buru Island	-	-	-	-	-	-	**0.35-0.4**	[14]
14	Buru Island	-	-	-	-	-	-	**8.05-10.23**	[15]
Marine biota: mollusc									
15	Ambon Bay	-	-	<0.001-0.47	0.34-0.77	0.25-0.36	<0.001-0.47	-	[5]
16	Ambon Bay	-	-	-	-	**3.908-5.490**	0.309-0.694	0.196-0.410	[12]
17	AIB	-	11.53-38.96	-	-	0.78-2.36	002-0.046	-	[16]
18	AIB	0.09-0.36	-	0.14-0.51	-	-	0.01	-	[17]
19	AIB	1.415 - **4.250**	-	18.145-48.040	-	-	0.315-1.415	-	[18]
20	Buru Island	-	-	-	-	-	-	0.15-**2.99**	[15]
Marine biota: fish									
27	Buru Island	-	-	-	-	-	-	0.001-**3.44**	[15]
Marine biota: crustacea									
28	Buru Island	-	-	-	-	-	-	0.12-**120.31**	[15]

continued on the next page

Table 1. Continued

No	Location	Cr	Fe	Cu	Zn	Pb	Cd	Hg	Ref
Marine biota: soft coral									
21	AOB	-	-	0.8975-2.7962	6.36-22.8025	1.4612-8.57	-	-	[19]
22	AOB	1.95-2.585	-	-	-	8.2-12	0.755-1.0662	-	[20]
Marine biota: seagrass									
23	Ambon Bay	8.61-18.29	-	-	-	14.2-37.5	13.18-17.43	-	[22]
24	Ambon Bay	15.7-22.65	-	-	-	14.76-20.88	15-27.35	-	[23]
25	AIB	-	-	-	-	2-20	-	-	[9]
Marine biota: sea urchin									
26	Ambon Island	-	-	-	-	-	0.30-1.95	-	[8]
Guidelines									
Sea Water		0.005	-	0.008	0.05	0.008	0.001	0.001	[6]
Marine sediment		81	-	34	150	46.7	1.2	0.15	[10]
Mollusc		2	-	70	1000	2.5	2	1	[5]
Fish and crustacean		-	-	-	-	0.5	-	0.5	[19]

However, the Cd and Pb concentration in Ambon Island tend to increase since 2011 (table 1). The study in 2011 reported the high Cd level was observed in four sites around Ambon Island (Latta, Liang, Latuhalat and Tial) (figure 1) [8]. These sites are less populated in Ambon Island. The study in 2014 also reported the higher Pb in Poka and Lateri [9]. These sites are dense population area in Ambon Inner Bay. These concentrations were above the aforementioned standard.

2.1.2 Marine Sediment

An initial study of heavy metals in marine sediment in Ambon Bay, the Banda Sea, was reported in 1983 [4] (table 1). This study indicated low concentrations of Cu, Zn, Pb, Cd and Hg. The concentrations observed were below the guideline issued by the National Oceanic and Atmospheric Administration Effect Range Low (NOAA ERL) [10]. The second study published in 1987 indicated Cd and Hg concentration increase. However, the concentrations of Cu, Zn, Pb observed were below the aforementioned guideline [11].

Almost two decades after the first publication about heavy metals in marine sediment in the Banda Sea, a study in Batu Gong, Hative Besar, Poka and Seri indicated higher concentrations of Pb (~36 times) and Hg (~28 times) than those reported in 1987 (see table 1) [12]. Pb and Hg levels in 2005 have exceeded the NOAA guideline. In contrast to increasing Pb and Hg concentrations, the level of Cd was lower than those observed in 1987.

An indication of higher Zn concentration (~5 times) than those reported in 1987 was also reported in 2009 [7]. This study also reported a lower Pb and Hg level than those observed in 2005. However, Pb, Zn, Cd and Hg levels found in this 2009 study were below the NOAA guideline. The heavy metal concentrations below the NOAA guideline's upper limit were also observed in Latta, Liang, Latuhalat and Tial [8] and Poka and Lateri [9].

The most updated information regarding heavy metals concentrations in the Ambon Bay was published in 2017. The reference [13] reported the level of Fe, Zn, Cu, Pb, Cd and Hg in marine surface sediment of Ambon Bay were higher than those measured in previous studies. Based on the quality standards for marine sediment under NOAA value, level of certain metals (Cu, Zn, Hg) measured were above the guideline. There was no guideline found for Fe [10]. This study indicated the strong contribution of heavy metals pollution from the urban waste, residential sources, farming industry, shipping activities and ship repair activity from dockyard around the Ambon Bay. The increasing trend of heavy metal on marine sediment on those studies can be expected that many pollutants accumulate in the sediments.

Recent studies [14, 15] reported the Hg contamination in marine sediments offshore Kayeli Bay Buru Island). The Hg concentrations observed on both studies were above the aforementioned standard for marine sediment. This Hg contamination was suspected to be originated from artisanal gold minings operating in this island. These studies were sampled in 2012 and 2013. The interval time between those study was 10-month period, but surprisingly the Hg concentration in 2013 has increased more than 20 times than those sampled in 2012.

2.1.3 Marine Biota

Investigations about heavy metals concentration in marine biota were done mostly on molluscs (6 studies), followed by studies on soft corals (2 studies), seagrass (3 studies), sea urchin (1 study), fish (1 study) and crustacean (1 study). Heavy metal studies on molluscs were done from 1987 to 2012.

The first study reported a low concentration of Pb, Cd, Cu, Zn on molluscs *Chicoreus* sp., *Conus quercinus* and *Strombus tuhuanu* [5] (table 1). The concentrations recorded were below the National Health and Medical Research Council (NHMRC) threshold value [5]. To our knowledge, the NHMRC was the only guideline widely cited for molluscs. However, online tracking of this reference did not yield satisfactory result nor any update of this guideline. The second study reported the Pb, Cd and Hg level on molluscs Neritidae (*Nerita albicilla, N. planospira, N. umlaasiama, N. picea, N. literata, N. maxima*) [12]. The level of Pb found has exceeded the NHMRC limit. However, the Cd and Hg concentration on those molluscs were below the NHMRC value. The third study reported the low concentration of Cd and Pb found on the mollusc *Anadara antiquata* [16]. The Cd and Pb observed were below the aforementioned guideline. In 2007, the level of Cu, Cd and Cr measured on molluscs *Gafrarium tumidum* [17] and *Soletellina biradiata* [18]. Higher concentrations of these heavy metals were found on the latter (table 1), notably the Cr has exceeded the NHMRC limit. This study also observed the level of Fe on mollusc *Gafrarium tumidum* but no guidelines exist about Fe concentration [5].

Five of the six previous studies about heavy metal contamination on mollusc were done in Ambon Island. The other one reported the Hg concentration from Kayeli Bay (Buru Island) [15]. The Hg concentration were measured on four mollusks species: *Anadara granosa, Cerethium* sp., *Nerita* sp. and *Terebralia palustris.* The highest Hg concentration was recorded on *Cerethium* sp., followed by *Nerita* sp., *A. Granosa* (raw), *T. palustris*, and *A. granosa* (cooked). The Hg concentration found on *Cerethium* sp. has exceeded the aforementioned guideline. Moreover, this study also reported the Hg concentration on 12 fishes and 3 crustaceans from Buru Island. The higest Hg concentration was found on a freshwater/estuarine species *Channa striata* followed by *Terapon theraps* (omnivorous), *Epinephelus fuscoguttatus* (benthic, carnivorous), *Equulites leuciscus* (benthic, carnivorous), *Lutjanus rufolineatus* (carnivorous), *Upeneus vittatus* (carnivorous), *Dussumieria* sp. (planktivorous), *E. fasciatus* (carnivorous), *Alectis ciliaris* (carnivorous), *Auxis rochei* (carnivorous), *Stolephorus indicus* (planktivorous), *Decapterus macrosoma* (a pelagic schooling fish common in deeper marine waters) [15]. The Hg concentrations on 5 species have exceeded the guideline issued by the Food Standards of Australia and New Zealand (FSANZ) [19]. The Hg concentration on crustacean was recorded on *Metapenaeus lysianassa*, Kepitang bakau (Indonesian name, possibly for *Scylla* spp. a type of crab species that live in the mangrove ecosystem), *Metapenaus* sp. The Hg concentration on the prawn *M. lysianassa* were higher more than 100 times than Hg observed on Kepitang bakau and *Metapenaus* sp. [15] and far exceeded the FSANZ value.

Several heavy metals concentration on soft corals (*Sarcophyton roseum* and *Symphylia agaricia*) were reported in 2007 [20] and 2008 [21]. Compared to other study, the Pb observed on soft coral in AOB were below the Pb observed on soft coral *Sinularia polydactyla* in Sulawesi (eastern Indonesia) [22].

There were three studies about heavy metals on seagrass in the Banda Sea. Two studies were conducted on seagrass *Enhalus acoroides*. The first one reported the Pb, Cd and Cr concentrations on its leaves [23]. The other one observed the Pb, Cd and Cr on roots [24]. The higher concentration of Pb, Cd and Cr were higher in the roots. The third study measured the Pb concentration on leaves, rhizomes and roots of *Thalassia hemprichii* [9]. The results showed a consistent pattern of accumulation (rhizomes> leaves> roots) at all sites [9].
The latest study on heavy metal concentration in the Banda Sea reported the low Cd concentration on sea urchin *Diadema setosum* from Ambon Bay in 2011 [8]. The Cd measured in Ambon Bay was lower than Cd level observed on the same organism sampled in a coral reef area in Singapore [25].

The activity in the harbour area of Ambon Island was reported as the probable cause of distorted growth and reproductive failure (imposex) in some of the gastropods sampled in Ambon Island [26]. These gastropods were taken from 9 out of 24 sampling sites at which whelks were present. The imposex was observed in three gastropods groups (*Thais kieneri, T. savignyi* and *Vasum turbinellus*) that were found in harbour area but was absent or in mild stage where there was no harbour related-activity in the sampling area.

2.2. Hydrocarbon

Both studies about hydrocarbon concentrations in the Banda Sea were conducted in Ambon Bay. The first study reported the hydrocarbon concentrations in Ambon Bay were at 22.3-25.9 $mg.l^{-1}$ [27], and these were higher than the limits of the national standard on marine water quality (Decree of the Indonesian Ministry of Environment no. 51/2004) [6]. The second study was conducted in 2009 [28], where it was reported that the hydrocarbon content was ten times lower than those observed in 1991. However, the hydrocarbon level observed in 1991 was above the aforementioned standard. Other than that, an initial study about hydrocarbon-degrading bacteria from Ambon Bay was published in 1997 [29].

2.3. Domestic Waste

The research since 1995 obviously mentioned the high risk of plastic debris in Ambon Island [26]. The study of marine litter along Ambon Island coasts (10 sites in Ambon Bay, 6 sites in South-east waters of Ambon Island and 5 sites in the Northeast area) showed that the mean density of marine litter in Ambon Island was at 4.6 $items.m^{-2}$. Compared to the other countries, the number of mean density of marine litter in Ambon Island is higher than Brazil (0.14 $items.m^{-2}$), Australia (0.04 $items.m^{-2}$), Mexico (1.5 $items.m^{-2}$) and Japan (3.41 $items.m^{-2}$) [30].

The highest density was found in Ambon Bay and the less contaminated was south-east coast sites [26]. There was an assumption that the local human population was the source of marine litter due to the significant correlation of their presence on the shore and the number of population in the adjacent coastal area. In this study, there was also assumption that the domestic waste in Ambon Bay was an indirect cause of heavy loads of the isopod ectoparasite *Renocila* sp. in the coral reef fish, *Abudefduf saxatilis* in Ambon Island. Indeed, it was assumed that there was a positive correlation between the increasing number of parasite load and the increasing number of organic waste in the coastal areas sampled. Furthermore, counting carried out on the floating litter on the surface of Ambon Bay indicated its highest density near the market of Ambon (Pasar Mardika) and Ambon Inner Bay. The density of the beach litter along the strandlines and the upper tidal between Ambon Inner Bay (Poka) and Ambon Outer Bay (Silale) was also compared. The beach litter found in Poka was eight times higher than in outer bay [31]. The density difference was due to the sizes of human population in respective area, where Poka was heavily populated than Silale.

The study of Institute for Marine Research and Observation Denpasar, Indonesian Ministry of Maritime Affairs and Fisheries [32] reported the presence of 5 000-6 000 particles of microplastics per km^2 in the Banda Sea. Compared to the other studies, microplastics density in the Banda Sea was higher than the mean density of floating litter in the North Sea (25-38 $items.km^{-2}$), West Hawaii (0.5

items.km^{-2}) South China Sea (0.3-16.9 items.km^{-2}), North Pacific (459 items.km^{-2}), Strait Malacca (579 items.km^{-2}) and global density of floating marine debris (0 to beyond 600 items.km^{-2}) [33].

3. Future perspective of marine pollution research in the Banda Sea

Nowadays, plastic debris has become a global issue. Plastic debris is one of the most challenging problems for marine environment because slowly degrading plastic wastes generate microplastic (particles smaller than 1 to 5 mm) which can be easily spread over long distances by wind-driven ocean surface layer circulation [34]. This waste is ubiquitous in the ocean remote shorelines, coastal waters, and the seabed of the deep ocean and also floats on the sea surface. The quantity of floating litter (plastic debris) observed in the open ocean in mid-ocean gyres represents a small fraction of the total input [35].

The presence of plastic debris in the marine environment generates ecological and socio-economic impact. Lots of studies have proved that microplastics have entered the organism's body such as on zooplankton [36], tuna [37], whale [38], corals [39], crab [40] by ingestion process. Marine plastic debris could affect also the reputation of the local tourism industry [34]. Due to the potential risk of marine plastic debris and microplastic in the Banda Sea, more studies about plastic and microplastic debris in the Banda Sea is urgently required to investigate the source, distribution, fate and impacts of marine microplastics in the Banda Sea. Adopting the research needs from UNEP program, we suggest conducting these topics for research in the Banda Sea:

a) Sources of plastic and microplastics
- Source of plastic: research can be conducted to quantify inputs from fisheries sector, aquaculture sector, tourism sector, atmospheric transport.
- Source of microplastic: research can assess the relative contribution of synthetic fibers; the relative contribution of vehicle tire fragments; the size, shape, and composition (polymer and additives) of microplastics from different sources; the input of resin pellets from the plastics production and river inputs; atmospheric inputs.

b) Distribution and fate
- Factors controlling degradation
- Presence, transport, and fate of plastics in the marine environment

c) Impacts
- Quantifying impacts on biota, especially commercial biota such as tuna
- Measuring the socio impacts
- Assessing the economic impact

Acknowledgment
The author is thankful to Johanis Lekalete for generating the maps and both anonymous reviewers for their contributions in improving this paper. The participation of CYM in the ISBSE 2017 was fully funded by Centre for Deep-Sea Research, LIPI.

References

[1] Van Riel P M 1934 The bottom configuration in relation to the flow of the bottom water (*The Snellius Expedition in the Eastern part of the Netherlands East Indies 1929-1930* vol II part 2 Chapter II) (Utrecht: Kemink) 63 p

[2] Ministry of Maritime Affairs and Fisheries 2014 The total capture fisheries production, 2010-2014 *Marine and Fisheries in Figures 2015* [Internet] accessed on Nov 24th, 2017 from http://statistik.kkp.go.id/sidatik-dev/Publikasi/src/kpda2015.pdf

[3] GESAMP (IMO/FAO/Unesco/WMO/WHO/IAEA/UN/UNEP Joint Group of Experts on the Scientific Aspects of Marine Pollution) 1991 Reducing environmental impacts of coastal aquaculture *Rep. Stud. GESAMP* **47** p 35

[4] Thayib S S and Razak H 1988 Observation on the bacteriological indicator, heavy metal and pesticide at Ambon Bay, Banten Bay and Jakarta Bay, *Perairan Indonesia: Biologi,*

Budidaya, Kualitas Perairan dan Oseanografi, pp 124–31 (in Indonesian with English abstract).

[5] Edward and Tarigan 1987 Pengamatan pendahuluan kadar Pb, Cd, Cu dan Zn dalam air dan biota di Teluk Ambon (*Teluk Ambon: Biologi, Perikanan, Oseanografi dan Geologi*) eds S Soemodihardjo, S Birowo and K Romimohtarto (Ambon: LIPI Press) pp 106-11

[6] Kementerian Lingkungan Hidup 2004 *Keputusan Menteri Lingkungan Hidup no. 51/2004 tentang baku mutu air laut* (Jakarta: Kementrian Lingkungan Hidup)

[7] Marasabessy M D, Edward, Valentin F L and Radjab A W 2009 Kandungan logam berat di Teluk Ambon dalam kaitannya dengan kehidupan biota laut *Lingkungan Tropis Edisi Khusus* pp 383 – 92

[8] Rumahlatu D 2011 Konsentrasi logam berat kadmium pada air, sedimen dan *Diadema setosum* (Echinodermata, Echinoidea) di Perairan Pulau Ambon *Jurnal Ilmu Kelautan* **16**(2) 78-85

[9] Tupan C I, Herawati E Y, Arfiati D and Aulanni'am 2014 Profile of Lead (Pb) heavy metal in water, sediment and seagrass *(Thalassia hemprichii)* in Ambon Island, Maluku, Indonesia *J. Bio. Env. Sci.* **5**(4) 65-73

[10] NOAA (National Oceanic and Atmospheric Administration) 1999 *Screening quick reference tables (SquiRTs)* [Internet] accessed Feb 1 2018 from http://response.restoration.noaa.gov/cpr/sediment/squirt/squirt.html

[11] Edward and Pulumahuny F S 1987 Kandungan Logam Berat Hg, Pb, Cd, Cu dan Zn dalam sedimen di Perairan Teluk Ambon *Jurnal Fakultas Perikanan* **11**(1) 24-30

[12] Leiwakabessy F 2005 *Logam berat di perairan pantai Pulau Ambon dan korelasinya dengan kerusakan cangkang, rasio seks, ukuran cangkang kepada individu dan indeks keragaman jenis siput Nerita (Neritidae: Gastropoda)* Disertation (Surabaya: Airlangga University) 221 p

[13] Manullang C Y, Lestari, Tapilatu Y and Arifin Z 2017 Assessment of Fe, Cu, Zn, Pb, Cd and Hg n Ambon Bay surface sediments *Mar. Res. Indonesia* **42**(2) 77-86

[14] Male Y T, Reichelt-Brushett A J, Pocock M and Nanlohy A 2013 Recent mercury contamination from artisanal gold mining on Buru Island, Indonesia–potential future risks to environmental health and food safety *Mar. Pollut. Bull.* **77**(1–2) 428–33

[15] Reichelt-Brushett A J, Stone J, Howe P, Thomas B, Clark M, Male Y, Nanlohy, Butcher P 2016 Geochemistry and mercury contamination in receiving environments of artisanal mining wastes and identified concerns for food safety *Environ. Res.* **152** 401-18 56

[16] Pentury R 2006 *Analisis Kadar Logam Pb, Cd dan Fe dalam kerang* Anadara antiquata Undergraduate thesis (Ambon: Pattimura University) 56 p

[17] Hattu S 2007 *Analisis Kadar Logam Berat Cu, Cd, Cr dalam daging kerang manis (*Gafrarium tumidum*) pada Perairan Teluk Ambon bagian dalam* Undergraduate thesis (Ambon: Pattimura University) 47 p

[18] Nuniary M J 2007 *Analisa kadar logam Cd, Cr dan Cu dalam daging kerang* Soletellina biradiata *pada perairan Teluk Ambon Dalam* Undergraduate thesis (Ambon: Pattimura University) 57 p

[19] FSANZ (Food Standards Australia and New Zealand) 2015 *Contaminants and Natural Toxicants* [Internet] accessed Jun 26 2018 from https://www.legislation.gov.au/Details/F2015C00052

[20] Manuputty N 2007 *Penggunaan karang lunak (*Sarcophyton roseum*) sebagai indikator pencemaran logam berat Pb, Zn dan Cu di Perairan Teluk Ambon Bagian Luar* Undegraduate thesis (Ambon: Pattimura University) 47 p

[21] Tuwanakotta N 2008 *Kadar logam berat Pb, Cd dan Cr dalam spons (*Symphylia agaricia*) di Perairan Teluk Ambon bagian luar* Undegraduate thesis (Ambon: Pattimura University) 48 p

[22] Kadir H, Samawi M F and Haris A 2013 Akumulasi logam berat Pb pada rangka dan polip karang lunak *Sinularia polydactyla Jurnal Ilmu Kelautan dan Perikanan* **23**(1) pp 1-7

[23] Sipahelut A 2009 *Penentuan kandungan logam berat Pb, Cd dan Cr pada daun lamun* Enhalus

acoroides *di Perairan Teluk Ambon* Undergraduate thesis (Ambon: Pattimura University) 50 p

[24] Laimeheriwa, R S 2009 *Kandungan logam berat Pb, Cd dan Cr pada akar lamun* Enhalus acoroides *di Perairan Teluk Ambon* Undergraduate thesis (Ambon: Pattimura University) 55 p

[25] Flammang P, Warnau M, Temara A, Lane D J W and Jangoux M 1997 Heavy metals in *Diadema setosum* (Echinodermata, Echinoidea) from Singapore coral reefs *J. Sea Res.* **38** 35-45

[26] Evans S M, Dawson M, Day M, Frid C L J, Gill L A and Porter J 1995 Domestic waste and TBT pollution in coastal areas of Ambon Island (eastern Indonesia) *Mar. Pollut. Bull.* **30** 109-15

[27] Edward and Hermiati E 1990 Kandungan Hidrokarbon di Teluk Ambon, Maluku (*Perairan Maluku dan Sekitarnya*) eds D P Praseno (Ambon: LIPI Press) pp 106-11

[28] Dalila S 2009 *Kandungan senyawa Hidrokarbon di Perairan Teluk Ambon* Undergraduate thesis (Ambon: Pattimura University) 57 p

[29] Sutiknowati L I 1997 Sebaran Kepadatan Bakteri Pemecah Minyak di Teluk Ambon *Seminar Kelautan LIPI–UNHAS* (Makassar: LON LIPI–UNHAS) pp 260–64

[30] Pszczol S S O and Creed J C 2007 Quantification and classification of marine litter on beaches along Armação dos Búzios, Rio de Janeiro, Brazil *J. Coast. Res.* **23**(2) 421-28

[31] Uneputty P and Evans S M 1997 The Impact of Plastic Debris on the Biota of Tidal Flats in Ambon Bay (Eastern Indonesia) *Mar. Environ. Res*. **44**(3) 233-42

[32] Yunanto A 2017 *Pencemaran Sampah Plastik Mikro Terus Meluas* [Pollution continue to spread] *Kompas Daily* June 10th 2017 p 14 (in Indonesian)

[33] Galgani F, Hanke G and Maes T 2015 Global Distribution, Composition and Abundance of Marine Litter (*Marine Anthropogenic Litter*) eds M Bergmann, L Gutow and M Klages (Bremerhaven: Springer) pp 29-56 https://doi.org/10.1007/978-3-319-16510-3

[34] Thevenon F, Carroll C and Sousa J 2014 *Plastic debris in the ocean: The characterization of marine plastics and their environmental impacts, situation analysis report*. ed F Thevenon et al (Gland, Switzerland: IUCN) 52 p

[35] UNEP 2016 *Marine plastic debris and microplastics – Global lessons and research to inspire action and guide policy change* [Internet] accessed on Aug 15 2017 from https://wedocs.unep.org

[36] Cole M, Lindeque P, Fileman E, Halsband C, Goodhead R, Moger J and Galloway T S 2013 Microplastic Ingestion by Zooplankton *Environ. Sci. Technol.* **47**(12) 6646–55. doi: 10.1021/es400663f

[37] Romeo T, Pietro B, Pedà C, Consoli P, Andaloro F and Fossi M C 2015 First evidence of presence of plastic debris in stomach of large pelagic fish in the Mediterranean Sea. *Mar. Pollut. Bull.* **95** 358–61

[38] Besseling E, Foekema E M, Van F J A, Leopold M F, Kühn S, Bravo R E L, Heße E, Mielke L, IJzer J, Kamminga P and Koelmans A A 2015 Microplastic in a macro filter feeder: Humpback whale *Megaptera novaeangliae Mar. Pollut. Bull.* **95** 248–52

[39] Hall N M, Berry K L E, Rintoul L and Hoogenboom M O 2015 Microplastic ingestion by scleractinian corals *Mar. Biol.* **162** 725–32

[40] Watts A J R, Lewis C, Goodhead R M, Beckett S J, Moger J, Tyler C R and Galloway T S 2014 Uptake and retention of microplastics by the Shore Crab *Carcinus maenas Environ. Sci. Technol.* **48**(15) 8823–30 doi: 10.1021/es501090e

Fish larval composition and distribution assessment in the northern waters of Ambon Island

Y Tapilatu[1], A H Sabandar[2,*], V Salawane[2,*] and A Syahailatua[1]

[1]Center for Deep-Sea Research, Indonesian Institute of Sciences (PPLD LIPI), Jl. Y. Syaranamual Guru-guru Poka, Ambon 97233, Indonesia
[2]Undergraduate Programme, Faculty of Biology, Satya Wacana Christian University, Jl. P. Diponegoro 52-60 Salatiga 50711, Indonesia
*Both co-authors contribute equally

yosmina.tapilatu@lipi.go.id

Abstract. The Banda Sea ecosystem has been highlighted in the past decade for its potential role in providing economically important fish stock for human consumption. However, literature studies indicated lack of references regarding fish larvae abundance and composition in coastal areas of islands impacted directly by the Banda Sea. The study covers fish larval survey carried out at the northern waters of Ambon Island in April 2017. The composition, distribution, and abundance of fish larvae were described based on five horizontal and one vertical sampling stations. Twenty-eight families were identified and two of them were the most abundant: Atherinidae (23% of the total larvae) and Apogonidae (21% of the total larvae). Most of the other families were known to be neritic and in particular were fishes commonly found to be associated with coral reef ecosystems. This finding indicated the profile of the northern waters area, which has extensive coral reef coverage, as well as estuary and mangrove area.

Keywords: Banda Sea, ichthyoplankton, coral reef

1. Introduction

The Banda Sea is known as one of Indonesian important fishing grounds of commercial fishes, such as yellowfin tuna, skipjack, and small pelagic fishes. Administratively, it is a part of Maluku Province of Indonesia. It is included in the Fisheries Management Area (FMA) 714 together with the Tolo Bay waters. The Maximum Sustainable Yield of small pelagic, demersal and reef fishes' of the Banda Sea is recently estimated at 124 477 tons per year [1]. Coastal waters of small islands in the Banda Sea were reported as fishing grounds for the aforementioned groups of fishes. However, catching activities vary according to the monsoon periods. Maluku Province Fisheries and Marine Office reported in 2007 that the local fishermen of the Ambon Island northern coastal area will carry out these activities during the southeastern monsoon season (May-September) [2].

Studies on fish larvae in this area were few and were carried out mostly in the oceanic area of the Banda Sea ecosystem (c.f [3,4]). However, there is not much information, if any, available regarding fish larvae abundance and composition in coastal areas of islands impacted directly by the Banda Sea, such as the Ambon Island. This is due to the difficulties encountered when identifying fish larvae collected directly from the sea, as specimens are prone to degradation due to handling during sampling and pre-treatment prior to identification. Whereas it is possible that larvae of commercial fishes such

Content from this work may be used under the terms of the Creative Commons Attribution 3.0 licence. Any further distribution of this work must maintain attribution to the author(s) and the title of the work, journal citation and DOI.
Published under licence by IOP Publishing Ltd

as yellowfin tuna could be found in the coastal area [2]. For this reason, it is necessary to have the initial information to confirm its presence in the northern coastal area of Ambon Island and to study the relationship between the yellowfin tuna larvae and its adult in the Banda Sea if possible.

The objective of this study was thus to provide updated information regarding fish larvae in this area. This information can be used by the Maluku Province government to sustainably plan the management of fisheries resources of the area.

2. Material and Methods

2.1. Sample collection

Samples were collected from the northern coastal area of Ambon Island (figure 1) at April 10-11, 2017. The sampling site was chosen because it was known locally for long time to be fishing grounds, notably for small pelagic and coral reef fishes. Sampling was carried out during the low tide period. As a consequence, the sampling was carried out approximately 20 to 50 meters seaward from the shoreline. A single bongo net with 500 μm mesh was used for this purpose. The diameter of the net opening was 1.5 m, and the length was 2.5 m. It was equipped with 5 kg weight. Sampling was carried out during the night (19.00-20.00 local time), with the net towed horizontally at the speed of 2 knots for about 15 minutes. At one of the stations (Station 4), the net was also towed vertically (nylon rope length 20 m). A traditional light emitting device was used to attract larvae to the net. Samples collected were obtained from the net by spraying and rinsing, and subsequently stored in plastic bottles filled with 10% formalin until transfer to the fish laboratory.

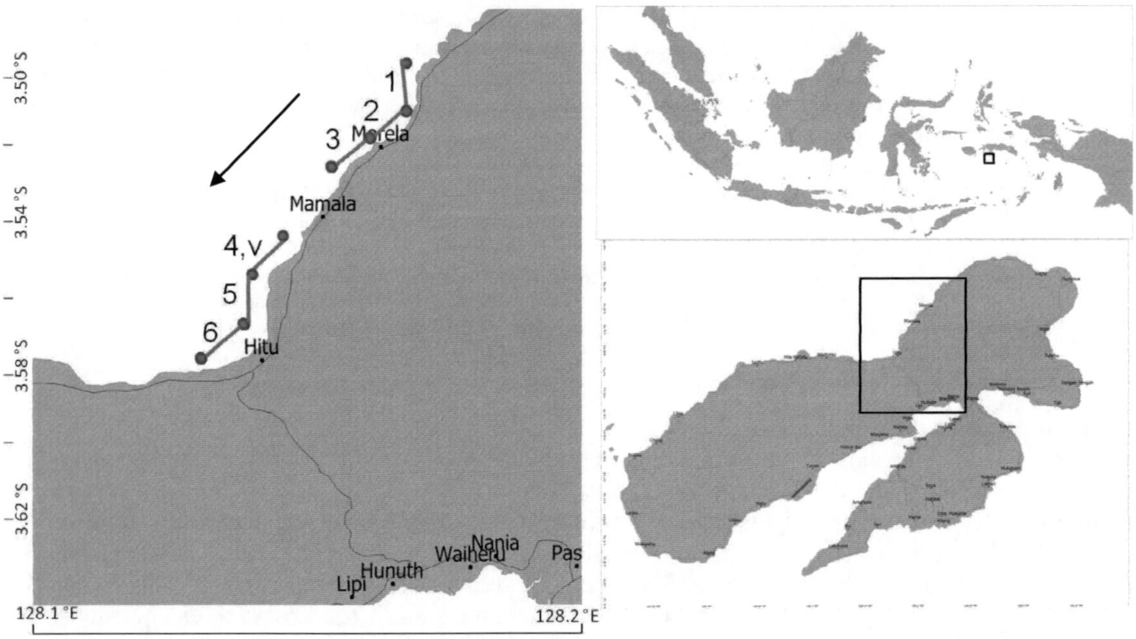

Figure 1. Map of Indonesia (inset, upper right), Ambon Island (inset, lower right) and sampling location (main, left). Blue dots indicate the starting and end points where sampling net was towed (2 knots, 20 minutes). Black arrow indicates the towing direction. Numbers (1-6) denote horizontal towing stations and v denote the location where the net larva was towed vertically.

2.2. Sample preservation and sorting

Samples obtained were diluted with tap water to reduce the preservative (formalin) concentration and sorted according to their family based on their morphology. Sorting was carried out manually. Small samples were sorted using a dissecting microscope (Nikon SMZ-645). Fish larvae were separated from the rest of the samples. Other organisms (copepods, phytoplankton, molluscs, and crustacean) were also sorted out and counted but not identified.

2.3. Sample counting, measurement, documentation, and identification

Samples were counted manually either visually or by using the dissecting microscope when their sizes were too small to be counted directly. Millimetre block paper was used to measure the total length (TL) of samples obtained as there was no ocular micrometer objective lens available. The pictures of fish larvae to be identified were taken manually using iPhone 6sPlus camera. High resolution pictures obtained were pooled according to stations and used as supporting data for identification. Samples were identified using several guidelines and identification books [5-8]. All samples were preserved after identification in 75% alcohol. Developmental stage of fish larvae was determined according to Kendall et al. [9] by examining the morphology of larvae collected during sampling, especially pigment pattern, larval body shape, absence or presence of armature on head bones and fin shape.

2.4. Calculation of fish larvae population density

Fish larvae population density was defined as the amount of larvae sampled per sampling area, using the following equation [10]:

$$T = 1000\, t / V \qquad (1)$$

with T is population density, expressed in number of fish per 1000 m^3 ; t is number of fish counted and V is volume of water filtered by net (1000 m^3). Due to the absence of flowmeter during the sampling period, the volume of water filtered by net was calculated by multiplying the area of net opening with the distance it was towed.

3. Results and Discussions

3.1. Sampling locations profile

The seabed in this coastal area is a shallow shelf from the intertidal area to about 50 meters seaward, and a steep slope is observed afterward. The sampling site was known as a fishing ground, notably for small pelagic and coral reef fishes. As has been described elsewhere in this proceeding [11], local fishermen used homemade bombs in the sampling area to fish (ca. 2005). Local authorities have intervened by educating them on the importance of keeping the coral reef ecosystem unharmed. This intervention resulted on the discontinuation of the blast fishing method in recent years. Tanjung Setan was popular locally as a recreational site, notably for swimming, snorkelling and diving. Fragments of branched coral found were mostly due to anchoring and tourism activities. Overall Morela coral reef ecosystem is in better condition compared to those in Mamala and Waipokol. Morela station was closed to the shore, where houses were built over the shoreline. Waitomu station was close to the river mouth where there was slight mangrove coverage at the shore.

The coral reef condition of this area in general was relatively good visually during the day on the sampling period. The depth was between 4 to 6 meters (table 1).

3.2. Total numbers of fish larvae obtained

Fish larvae were obtained in all stations, both horizontal and vertical. The number of larvae per collection ranged from 4 to 94, averaging 23 larvae (actual counts) per sampling station. They were largely outnumbered by other organisms (copepods, phytoplankton, mollusc and crustaceans) total count (5 072 samples) (figure 2). The highest percentage (18% of total sample obtained in the station) of fish larvae number compared with the other organisms was observed in Station 1 (Tanjung Setan). Nevertheless, in terms of actual count, the number of fish larvae obtained in station 1 was second the least (7 specimens only) compared to all the other stations.

Table 1. Sampling locations profiles.

Stations	Location	Position coordinates (Latitude, Longitude)	Approximate depth (m)	Dominant substrate
1	Tanjung Setan	3°31′00.0″ S, 128°13′23.4″ E	6	Coral reef
2	Morela	3°33′16.5″ S, 128°11′26.1″ E	6	Coral reef
3	Mamala	3°33′13.1″ S, 128°11′27.8″ E	4	Coral reef
4	Waitomu	3°35′08″ S, 128°06′44.0″ E	4	Rocky
v	Waitomu-vertical			
5	Wakal	3°35′14.9″ S, 128°09′16.8″ E	6	Sand
6	Waipokol	3°34′56.7″ S, 128°10′20.4″ E	6	Sand

The total number of fish larvae sampled was 230 (actual counts), including larvae that were too poorly preserved (disintegrated) to be identified (20 larvae). However, the highest count of sample (94 fish larvae) was found on station 4 (Waitomu). This station was situated at the river mouth that goes directly to the coastal area, and heavily impacted by household wastes. Station 4 also had the total highest count (1 827) of other organisms –predominantly molluscs (63%) - compared to all the other stations. It seems that the river mouth was rich of nutrients due to the dumped household wastes, which explain the high abundance of organisms obtained during net towing.

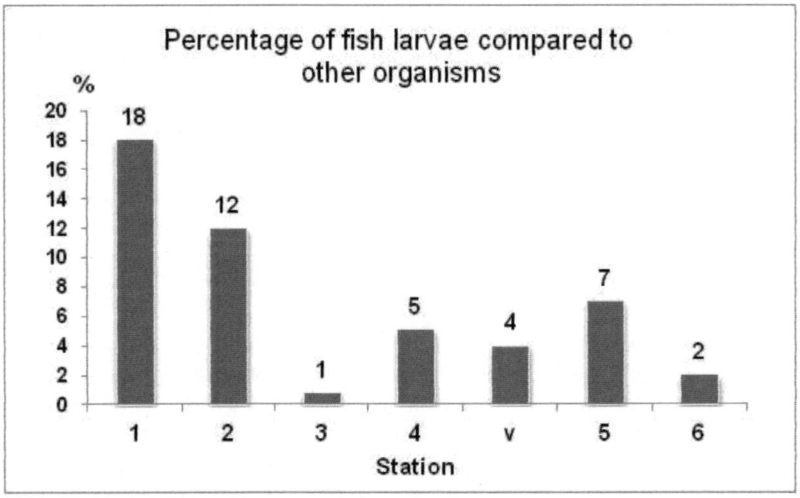

Figure 2. Percentage of fish larvae compared to other organisms in total samples obtained in each stations, horizontal (1-6) and vertical.

3.3. *Larval fish composition*

Larval fish composition obtained in this survey is summarized by family in Table 2. Larvae of more than 20 families were listed, but those of six families contributed to 66% of the total population density. Most of the larvae obtained were on the pre-flexion to flexion phases of larval development. The discussion here will be focussed on these six families.

The Atherinids (Silversides) larvae (22% of the total) were the dominant group with 84% occurrence of the total stations. They were found in all horizontal stations but none on the vertical one. This confirms their particular features of small, silvery schooling fishes that are closely associated with surface waters during all stages of life. Some species have larvae with dorsal pigment on the midbrain, short snout and terminal small mouth with large, round to ovoid, eyes [6]. Atherinids sampled were mostly between pre-flexion to flexion stage, as observed from their morphological features, particularly the incomplete growth of fins (see for ex. Fig. 3a). Their TLs were between 1 to 4.5 mm. About 46% of this family (from Atherinids population density) was found in station 4. This was explained by the high availability of their food (phytoplankton) in station 4, which reached 27% of total phytoplankton obtained from all stations.

Tabel 2. Population density, range of TL and larval stage of fish larvae obtained per station based on family. 1-6: horizontal towing, v: vertical towing at station 4. NR: Not relevant

No	Family	Population density (ind.1000 m^{-3}) per station							Total	TL range (mm)	Larval stage
		1	2	3	4	v	5	6			
1	Ambassidae	0	1	0	1	0	0	0	2	2.5-3.5	Flexion
2	Apogonidae	1	1	0	9	10	6	3	27	2.5-14	Pre-flexion to juvenile
3	Atherinidae	2	2	2	15	0	9	2	30	1-4.5	Pre-flexion to flexion
4	Aulostomidae	0	0	0	0	0	0	1	0	6	Pre-flexion
5	Berycidae	1	0	1	0	0	0	0	2	2-3	Pre-flexion
6	Callionymidae	0	0	0	0	0	1	0	1	4	Pre-flexion
7	Carangidae	0	0	0	1	0	0	0	1	3	Pre-flexion
8	Clupeidae	1	10	0	4	0	0	0	15	2.5-27	Pre-flexion
9	Dactylopteridae	0	0	0	1	0	0	0	1	3	Pre-flexion
10	Diodontidae	0	0	0	0	0	1	0	1	1	Flexion
11	Engraulidae	0	0	0	1	0	2	4	7	1.5-4	Pre-flexion
12	Exotidae	0	0	0	0	0	1	0	1	3	Pre-flexion
13	Gerreidae	0	0	0	1	1	2	0	4	3-18	Pre-flexion to juvenile
14	Gobiidae	0	0	0	0	0	1	0	1	6	Pre-flexion
15	Holocentridae	0	0	0	5	0	0	0	5	3	Pre-flexion
16	Labridae	0	0	0	1	0	1	0	2	2.5-5	Pre-flexion
17	Lutjanidae	0	0	0	3	0	2	0	5	3-5	Pre-flexion
18	Leiognathidae	0	2	0	2	0	0	0	4	3-4	Pre-flexion
19	Monacandthidae	0	0	0	1	0	0	0	1	4	Pre-flexion
20	Mullidae	0	1	0	1	0	1	0	3	2.4-5	Pre-flexion
21	Mugilidae	0	0	0	1	0	0	0	1	2.5	Pre-flexion
22	Plesiopidae	0	0	0	2	0	1	0	3	3	Pre-flexion
23	Pomacentridae	0	1	0	0	0	1	0	2	2.5-4	Pre-flexion
24	Serranidae	0	1	0	1	0	0	0	2	2.5-5	Pre-flexion
25	Sillagonidae	0	0	0	1	0	2	1	3	3-4	Pre-flexion
26	Synodontidae	0	1	0	0	0	1	0	2	2.5-5	Pre-flexion
27	Triodontidae	0	1	0	0	0	0	0	1	17-27	Pre-flexion
28	Triacanthidae	0	0	0	1	0	1	0	2	2-4	Pre-flexion
29	Disintegrated larvae	0	0	0	9	0	1	3	10	NR	NR
Total fish larvae (ind. 1000 m^{-3})		5	21	3	61	11	34	5	135		

The Apogonids (Cardinal fishes) were the second to have the highest population density (20% of total larvae). Fishes in this family are small, usually reddish, carnivorous and nocturnal living around coral reefs and other habitat, with larvae characterised by spotted lines on cheek bones and caudal fin, wide jaw bones and separated dorsal fins [6]. Most of them were already on the juvenile phase, indicated by their size (13-19 mm) that exceeded the other familial group. Its occurrence in large number at the vertical station could due to its nocturnal life style, as the sampling was carried out at night. They had about 84% occurrence of the total stations. However, contrary to the Atherinids ones, the Apogonids were found in all stations but in station 3. Despite the characteristic of this station (coral reef environment), samples from this station were the lowest in numbers, compared to the other stations.

Fish larvae belonging to the family Clupeidae ranked third in abundance with 9% of the total larvae taken in about 43% of the stations. They are small to moderate, silvery, planktivorous, schooling fishes of great commercial importance, with members such as herrings, sardinellas and sardines fishes [6]. About 69% of total Clupeids larvae were mostly found in station 2, an area predominantly covered by coral reefs, with many floating nets on site. Their TLs were the most varied, ranging from 2.5 to 27 mm.

Engraulidae (anchovies) larvae ranked fourth and exceeded 5% of the total population density, with 43% occurrence of the total stations. They are silvery, marine planktivores, schooling fishes of small to moderate size [6] (larvae sampled TL range: 1.5-4.5 mm). Commonly used as baits for big-pelagic fishing activities in Ambon waters and surrounding areas, their larvae are characterized by its very elongated body form [6] (figure 3d).

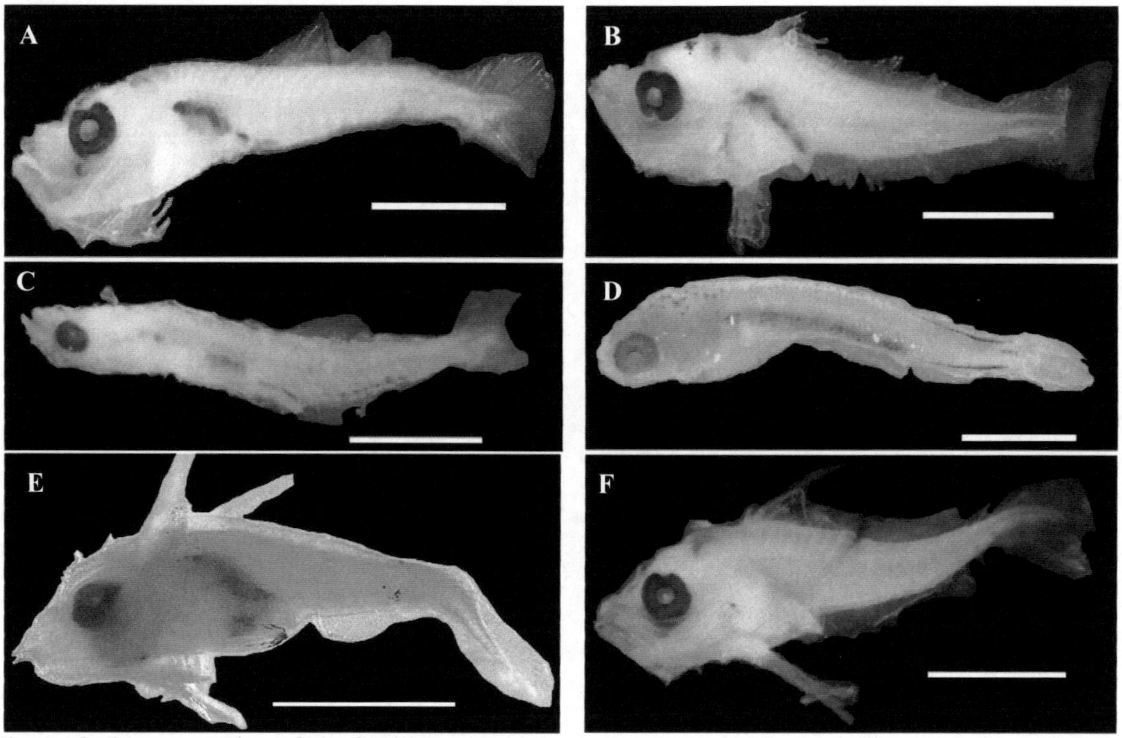

Figure 3. Representative of six families with highest occurrence during sampling: a. Atherinidae, b. Apogonidae, c. Clupeidae, d. Engraulidae, e. Holocentridae and f. Lutjanidae. Scale bar: 1 mm.

Holocentridae (Squirrelfishes) and Lutjanidae (Snappers and Fusiliers) larvae ranked fifth and each exceeded 3% of the population density with occurrences of 14% (Holocentridae) and 28% (Lutjanidae) of the total stations. Holocentrids are small to moderate in size, nocturnal, large-eyed, usually red fishes that shelter in the recesses of coral reefs during the day, with some species are known to be commercially important food fishes [6]. In contrast, Lutjanids are medium to large, carnivorous fishes found in various habitats such as coral reefs, mangroves, sandy bottoms and deep, rocky reefs, with many economically important fishes [6].

One third of the larvae obtained belong to the other 22 families, with occurrence of more than 50% of these families in station 4, compared to other stations. Their total counts ranging from 1 (e.g. Aulostomidae, Callyionimidae, Dactyloptiridae) to 6 (Gerreidae, Leiognathidae) samples only.

Some of the larvae sampled belonged to the families of ornamental tropical fishes reported previously to be found in Ambon Island coral reef ecosystem [12], such as Pomacentridae, Labridae, Serranidae, Monacanthidae and Gobiidae. However, the identification carried out in this study was only to family level, and it was thus inconclusive, whether the larvae obtained were indeed from specific species that are known to be commercially traded as ornamental fishes or not.

Almost 9% of larvae obtained were too poorly preserved (disintegrated) to be identified, and some others did not even have intact gas bladder, which made the identification work difficult. However this is the common problem faced when sampling fish larvae from the sea, as the degradation due to pre-treatment and/or preservation phase were prone to happen and thus might damage the samples. This was comparable to the damaged specimens obtained during the EASTROPAC I cruise, for example, that reached 11% of the total specimens collected [13].

No scombrid larva was found on this sampling site. This was possibly because the sampling site is not a breeding and/or nursing ground for this family. Fishes may be constrained in the location of spawning sites by their physiology, especially during early life-history stages ([14] and references therein). Other than that, it could also due to the environmental parameters such as sea surface temperature that was not conducive for Scombridae fishes breeding. Scombrid fishes such as yellowfin tuna breed on specific temperature (25-29°C), whereas the sampling took place in the middle of transition period from west monsoon to east monsoon, where average sea surface temperatures range from 30-31°C. Unfortunately no temperature measurement was carried out during the sampling period to justify this explanation. Other than that, it was also probably due to the distance from the coastal area, as well as the period when the sampling was carried out. An ichtyoplankton survey carried out in the off-coast area of the Savu Sea, for instance, collected some tuna larvae. The sampling was carried out in October and November 2016 using the oblique tow approach, at a location where the depth was at 80-120 m (Simanjuntak C, personal communication). Sampling on east monsoon period at nearby location but on deeper water column is thus deemed necessary to confirm these hypotheses of scombrid larvae absence.

In regard with the local fisheries management, ornamental and/or fishes caught for food are prone to overfishing by local fishermen. It is thus highly recommended that the fishing activity should be followed up closely by the relevant authorities to ensure the sustainability of the fisheries resources in this area.

4. Conclusion

Samples identified in this study indicated the profile of Ambon Island northern coastal waters, which are dominated by coral reef, mangrove and estuary ecosystems. Larval groups with the highest occurrence belonged to Atherinidae and Apogonidae, two coastal associated fishes with different features. The other families found during this study were mainly associated with coral reefs ecosystem. No scombrid larva was found, and this could due to the lack of supporting environmental parameters for its spawning on the sampling period and/or the sampling locations that were not typical of this larval group spawning and/or nursery ground.

Acknowledgment

This research was supported financially by LIPI's internal funding for research and development, as a part of the research project on sustainable utilization of marine bioresources in Ambon Island northern coastal waters. The authors wish to thank two anonymous reviewers who have helped on the improvement of this manuscript substantially. We also want to thank Abdul Wahab Radjab, the coordinator of the project, to facilitate the sampling trip, and La Pay, Jance Hehuwat and Tamara Ginzel for their assistance during sampling, preservation and identification of the samples. This is contribution number 0009 from Marine Microbiology and Biotechnology Laboratory, Center for Deep-Sea Research LIPI.

References

[1] Suman A, Wudianto, Sumiono B, Badrudin, Nugroho D, Merta GS, et al. 2014 *Potensi Dan Tingkat Pemanfaatan Sumberdaya Ikan Di Wilayah Pengelolaan Perikanan Republik Indonesia (WPP RI)* eds A Suman, Wudianto, B Sumiono, Badrudin, D Nugroho (Jakarta: Ref Graphika) 224 p

[2] Noija D, Martasuganda S, Murdiyanto B and Taurusman A 2014 Potensi dan Tingkat Pemanfaatan Sumberdaya Ikan Demersal di Perairan Pulau Ambon Provinsi Maluku *Jurnal Teknologi Perikanan dan Kelautan* **5**(1) 55–64

[3] Soewito and Schalk P H 1990 Spatial and Seasonal Patterns in Fish Larvae Distribution in the Banda Sea (Indonesia) *Neth. J. Sea Res.* **25**(4) 591–600

[4] Taufik M, Suwarso and Nurwiyanto 2005 Distribusi Kelimpahan Ichthyoplankton di Teluk Tomini dan Laut Banda *Jurnal Penelitian Perikanan Indonesia* **11**(6) 73–81
[5] Leis J and Carson-Ewart B 2000 *The larvae of Indo-Pacific coastal fishes: an identification guide to marine fish larvae (Fauna Malesiana Handbook 2)* 1st ed (Leiden: Brill) 850pp
[6] Leis J M and Carson-Ewart B M 2004 *The larvae of Indo-Pacific coastal fishes: an identification guide to marine fish larvae (Fauna Malesiana Handbook 2)* 2nd ed (Leiden: Brill) 850pp
[7] Mwaluma J M, Kaunda Arara B and Strydom NA 2014*A Guide to Commonly Occurring Larval Stages of Fishes in Kenyan Coastal Waters* [Internet].WIOMSA. Western Indian Ocean Marine Science Association 73 p accessed Aug 1 2017 from http://www.wiomsa.org/
[8] Termvidchakorn A, Hortle K G 2013 A guide to larvae and juveniles of some common fish species from the Mekong River Basin *MRC Tech Paper* 38 248pp
[9] Kendall A W, Ahlstrom E H, Moser H G. 1984 Early Life History Stages of Fishes and Their Characters (*Ontogeny and Systematics of Fishes*) eds H G Moser, W J Richards, D M Cohen, M P Fahay, A W Kendall Jr and S L Richardson (Kansas: Am. Soc. Ichthyol. Herpetol. Spec.) pp 11–22
[10] Servidad-Bacordo R A, Dickson L, Nepomuceno L and R Ramiscal 2012 Composition, Distribution and Abundance of Fish Eggs and Larvae in the Philippine Pacific seaboard and Celebes Sea with Focus on Tuna Larvae (Family: Scombridae) *WCPFC Sci. Comm. 8th Reg. Sess*. WCPFC-SC8-2012/SA-IP-03 25 pp
[11] Tapilatu Y 2018 Assessment of tunicates found in the northern waters of Ambon Island and its rearing requirements in a laboratory setting *IOP Conf. Ser.: Earth Environ. Sci.* **184** 012010
[12] Hukom F and Syahailatua A1995 *Distribusi dan kelimpahan relatif ikan hias laut di perairan Pulau Ambon dan sekitarnya* [Internet] 15pp accessed Aug 1 2017 from http://www.coremap.or.id/downloads/1274.pdf
[13] Ahlstrom EH 1971 Kinds and abundance of fish larvae in the eastern tropical Pacific, based on collections made on EASTROPAC I *Fish. Bull.* **69**(1) 3–77
[14] Ciannelli L, Bailey K and Olsen EM 2015 Evolutionary and ecological constraints of fish spawning habitats *ISE J. Mar. Sci.* **72**(2) 285–96

Reef fish stocks assessment around the islands in the Banda Sea for supporting local community livelihood

S Wouthuyzen[1], F D Hukom[2], P Makatipu[3] and D Pelasula[4]

[1]LPKSDMO - LIPI, Pari Island, Jl. Raden Saleh 43, Jakarta 10330, Indonesia
[2]Research Centre for Oceanography, Indonesian Institute of Sciences (P2O - LIPI), JL. Pasir Putih 1, Ancol Timur, Jakarta 14430, Indonesia
[3]LKBLB - LIPI, Bitung, Jl. Tandurusa, Aertambaga, Bitung 95527, Indonesia
[4]Center for Deep-Sea Research, Indonesian Institute of Sciences (PPLD LIPI), Jl. Y. Syaranamual Guru-guru Poka, Ambon 97233, Indonesia

sam.wouthuyzen@gmail.com

Abstract. Located in the Indonesian fisheries management area (WPP) 714, the Banda Sea has a high potential of fisheries resources. However, information about reef fish resources is limited. This study aims to assess the Banda Sea reef fish diversity, abundance, and stocks. The diversity, abundance, and density of three reef fish groups (indicator, target, and major) were assessed using underwater visual census (UVC) from 287 sites. A stock assessment was done by multiplying fish habitat areas and fish density. There were 287, 42, and 301 species of target, indicator, and major groups recorded respectively. High abundances were found in Banda Islands and the Southeast region, while the lowest abundance was in the Kendari waters. The average density of indicator and major groups were 0.251 and 4.519 individuals.m^{-2}, with estimated stocks of $2{,}883\times10^{6}$ and $51{,}899\times10^{6}$ individuals (ind.) respectively. The density of target group fish which has economic values was 1.976 ind.m^{-2}, equivalent to the biomass of 599.6 ton.km^{-2} with stocks of 4,214,468 tons. Despite huge reef fish resources supporting the local livelihood, the stock tended to decrease in 2009-2012 and 2015-2016 in Flores and Banda Islands. Therefore, conservation and sustainable use of reef fishes of the Banda Sea is required.

Keywords: Reef bathymetry, Reef fishes, Diversity & Abundance, Stocks, Banda Sea

1. Introduction

The Banda Sea is a portion of the western South Pacific Ocean and one of the inner seas in the Indonesian territory. It is bounded by the islands of the Maluku Province (Watubela, Kai, Tanimbar, Wetar, Babar, Islands) on the east, East Nusa Tenggara (NTT) Province (Lembata and Flores Islands) on the south; the islands of Southeast Sulawesi Province (Wakatobi Islands and Islands around Kendari waters) on the west, and Sulu, Buru and Seram Islands of the North Maluku and Maluku Province on the north [1]. Two deep basins are located in this sea; the north and south basin with the depth of 5,800 m and 5,400 m respectively. A volcanic ridge further divides the southern South Banda Basin from the Weber Basin where the deepest area reaches 7,440 m. An active volcano, Mount Api rises from the floor of the southern basin at depth of 4,500 m to 670 m above sea level. The clear waters surrounding many islands provide a perfect habitat for spectacular coral reefs [1].

Ekman in 1953 divided the Banda Sea into two regions on the basis of the realm of marine biogeography, the West Indo Pacific Regions and the East Atlanto Pacific. The West Indo Pacific is a huge areas spread from Southeast Asia to Papua New Guinea (PNG) consisted thousands of coral islands that have the highest of marine biodiversity [2]. On the other hand, Spalding developed a global system that divided the marine and coastal ecosystems in the world into 232 ecoregions. Those

Content from this work may be used under the terms of the Creative Commons Attribution 3.0 licence. Any further distribution of this work must maintain attribution to the author(s) and the title of the work, journal citation and DOI.
Published under licence by IOP Publishing Ltd

ecoregions were divided based on the homogeneity of marine biodiversity and the presence of the small ecosystems, or the existence of difference oceanographic or topography conditions [3]. Regarding those criteria, Indonesian waters were divided into 12 ecoregions in which the Banda Sea was grouped into ecoregion no.131 [4]. The Banda Sea is surrounded by other inner seas e.g. the Flores Sea (west side), Sawu Sea (southwest), Timor Sea (south), Arafura Sea (southeast), Seram and Maluku Seas (north). Furthermore, the Ministry of Marine Affairs and Fisheries (MMAF) manages the Banda Sea under two regulations: PerMen 01 2009 and PerMen 18 2014 where it was categorized as Fisheries Management Area (WPP) no 714.

There are about 533 coral species found in the Banda Sea, making it a part of the coral triangle system (coral diversity > 500 species) [5], as well as reef fishes that comprised about 1,760 species in which 3 species are endemics [4]. This accounts for about 30 % out of 6,000 of the world's reef fishes, or about 44% from the total of 4,050 reef fish species in the Indo-Pacific region [6,7]. Since the Banda Sea is also located in the eastern part of Indonesia, it is thought to have the highest diversity of reef fish in the world for the families of Pomacentridae (Damselfishes) and Chaetodontidae (Butterflyfishes) with 128 species and 83 species, respectively [8]. This high diversity of Banda Sea was very contrast with the total of reef fishes in the Hawaian waters, which has only about 420 species [9].

The richness of coral reef ecosystems provides direct or indirect goods and environmental services for the local community, especially in fisheries and marine tourism sectors [10]. However, the information about the coral reef resources, particularly the reef fishes, in the Banda Sea are still lacking since the data and information are very limited [11]. Those data, if any, are scattered over a vast geographic area of the Banda Sea. Therefore, it is difficult to establish a sustainable reef fish resources management.

Concerning the reef fish management, several obstacles exist, i.e. the lack of information on stocks of various reef fishes in the Indonesian fisheries statistics compared to pelagic and demersal fish. Thus, the existing reef fish stock data is inadequate to be used for management purposes [12]. Furthermore, apart from the available methods in estimating stock pelagic and demersal fish, there is still no practical standard method available to estimating stocks of the reef fish [13] rather than underwater visual census (UVC) which is time-consuming and requires expertise to identify species of reef fish during conducting an underwater census. In addition, most reefs are located in remote areas and in shallow waters with irregular shapes, making it difficult to access and perform acoustic surveys to estimate the stocks of reef fish [14].

Considering the importance of supporting the livelihood of local communities, this resource is facing threats of overexploitation and unsustainable fishing practices such as blasting/poisoning [15-17]. Therefore, stock assessment and sustainable management on reef fish in all Indonesian waters using effective and efficient ways are important and must be prioritized. This study aims to describe the diversity, abundance and to estimate the stocks of reef fish of the Banda Sea (WPP 714) as a part of supporting sustainable management of reef fish resources.

2. Methodology

Fish stock is defined as a group of fish belongs to a taxonomy unit such as species or sub-species that occupy in a certain geography region [18]. Fish stock assessment is the activity to get information on how much the fish stocks in a certain water body. This important information is then used to give guidance to the decision maker for managing and utilizing the fish resources properly [18-20]. Principally, reef fish stock assessment is simple, although there is no standard method available for assessing the stock rather than UVC [13]. Two parameters are needed, the estimated of shallow areas (A, ha or km^2) of reef fishes habitat and density of reef fish (D, $ind.m^{-2}$ or $ton.km^{-2}$) [13].

2.1. Study sites and shallow water bathymetry mapping of coral reefs

This study was conducted in 9 locations surrounding the Banda Sea that were partitioned into 8 regions i.e. Ambalau (1) and Lease (2) Islands as north region; Watubela Islands (3) as northeast region; Banda Islands (4) as mid-region, Kai Kecil Islands (5) as east region; Southwest Maluku Islands (6) as south region-Maluku; Lembata, Flores, Alor Island (7) as south region - NTT, Wakatobi Islands (8) as southwest region and the islands in the Kendari waters (9) as west region (figure 1).

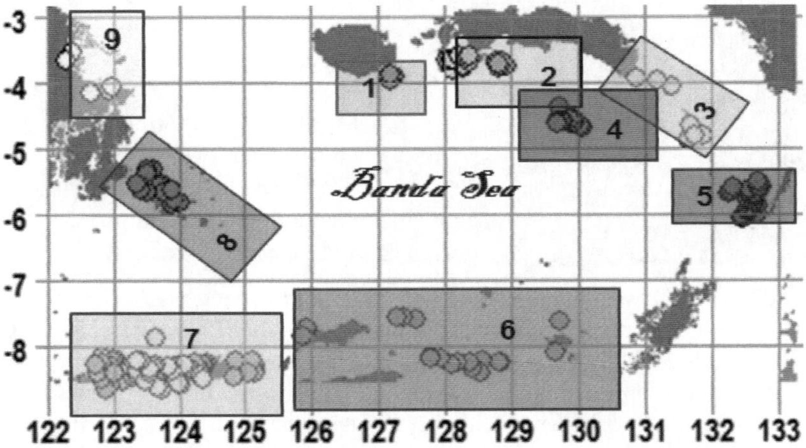

Figure 1. Map of the Banda Sea including the islands where reef fish stock assessment were conducted. Numbers denote locations

The first step in reef fish stock assessment was the determination of the shallow water on the range 0-20 m derived from bathymetry map to define the coral reef areas as the habitat of reef fishes. The bathymetry of shallow coral reefs can be mapped using remote sensing technique by utilizing satellite data such as Landsat (resolution 30 × 30 m). However, it needs at least 30 scenes of Landsat data (figure 2), which is impractical and time-consuming for deriving shallow reef bathymetry. As the alternative, available bathymetry data from the general bathymetric chart of the oceans (GEBCO) provided by the British Oceanographic Data Center (BODC) were used. This data has a resolution of 30 arc-second grid (926 m × 926 m) [21]. The image processing software of IDRISI Sevla (Ver. 17) was used to extract the bathymetry values. Depth range of 0-20 m was selected by assuming most of the reef fishes were abundance at that depth. The size area (A, km^2) on that depth was calculated afterward.

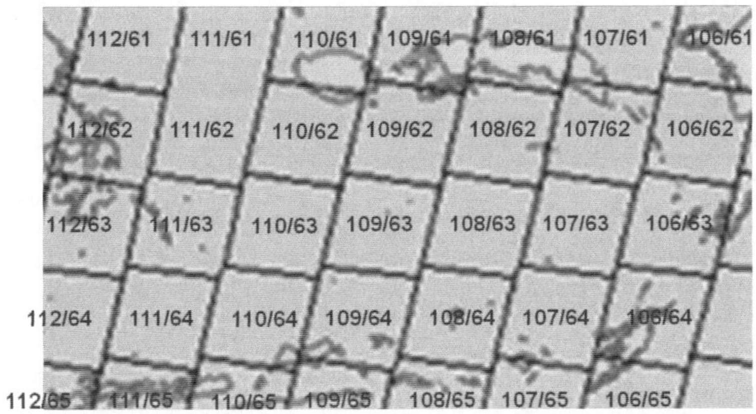

Figure 2. World References System (WRS) for of path and row number of the Landsat imagery for the Banda Sea.

2.2. Reef fish diversity, abundance, and density estimation
The second step was the estimation of the density (D) of reef fishes using UVC method (figure 3). This method was developed and introduced in the Asian-Australia project in 1986 [22] and applied for the first time in Indonesian waters. This method has become a standard operating procedure (SOP) for monitoring reef fishes in the Great Barrier Reef, Australia [23] and then well used in some other tropical countries.

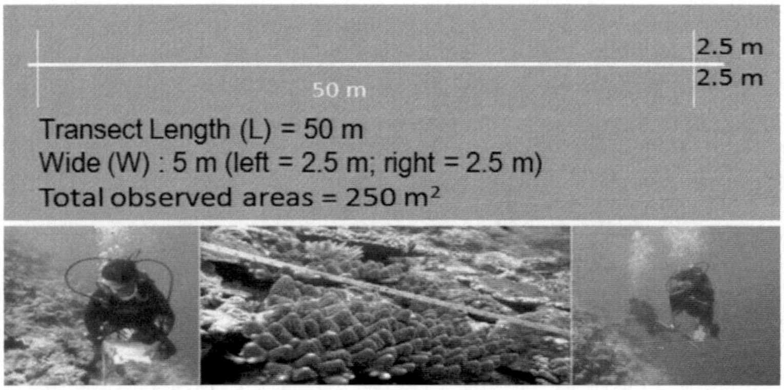

Figure 3. Illustration of Underwater Visual Census (UVC) method for estimating the biodiversity, abundance and density of reef fishes.

The UVC was performed simultaneously with the coral observation. A 50 m long plastic ruler tape was placed parallel to the coastline on the reef slope at depths between 3 and 10 m [22, 23]. The person who performing census dived using SCUBA equipment along this transect line and recorded all fish at the distance of 2.5 m to the left and 2.5 m to the right. This observation generated an observation area of 250 m^2 (figure 3). Reef fishes were identified using several fish identification books [24-28]. Each species was classified into three different groups; target, indicators and major groups [29]. The density of each species (D, $ind.m^{-2}$) was counted. Target group fishes are the consumable fishes, which have economic values, so they become a target by the fishers; Indicator group fishes are the fishes that strongly associated with coral reefs. All butterfly fishes (Family Chaetodontidae) belong to this group. The major group, the fishes that are not included in those two groups have small size, beautiful colour and well known as ornamental fishes [29].

2.3. *Data collection and analysis*

Two types of data were used; first, data collected from the field (primary data), and secondly, data collected from various sources, such as research reports originated from individual and/or NGOs or published journals (secondary data). Most secondary data used the same UVC method even though some sources needed to be standardized or adjusted or recalculated in order to have the same format as primary data format. Variables used in the primary/secondary data were reef fish diversity (number of species), abundance (number of fishes in each species), the density of each species (number of fish per area), date, and location per coordinates of census.

The standing stock (S) of each group of reef fish was estimated using the equation:

$$S = A \times D \quad (1)$$

where stock (S) is the total number of individual fish within each species of indicatior and major groups (individual), A is the size area (m^2 or km^2) and D is the density of each species ($ind.m^{-2}$). For the target group, since all species in this group have economic value, then S was expressed in the total biomass, rather than numbers of individual fish. The weight-length relationship:

$$W = a \times TL^b \quad (2)$$

was used for the conversion, where W denotes fish biomass (tons) and TL is the fish total length in cm, while a and b are fish growth coefficients obtained from available references [30-35]. Equation 2 was multiplied afterwards with the total number of specific species counted during the census to obtain the total biomass of the aforementionned species. The fish length was estimated mostly using the following equation:

$$TL = (TL_{min} + TL_{max})/2 \quad (3)$$

where TL_{min} is the minimum TL and TL_{max} is the maximum one. If there is no information available on TL_{min} and/or TL_{max} for specific species to be calculated using equation 3, we used the information of common/mean TL of fishes from previous literatures (eg. [30,33]), as well as the information of TL_{max} [33] to estimate the species TL (TL_s):

$$TL_s = 0.65 \times TL_{max} \quad (4)$$

The multiplication factor used in the equation 4 (0.65) was based on approximation and field experience.

In order to ensure whether any influence of the upwelling phenomenon to the stock of reef fishes, both data chlorophyll-a concentration and sea surface temperature (SST) derived from Aqua-MODIS satellites images (2003-2017) were used in the analysis. These data were obtained from Giovanni online data system web (Geospatial Interactive Online Visualization ANd aNalysis Infrastructure) developed and maintained by the NASA GES DISC.

3. Results and Discussion

3.1. Bathymetry map of the Banda Sea

Figure 4 shows the bathymetry map of the Banda Sea derived from GEBCO and was processed using IDRISI Selva software. The depth of the Banda Sea ranges from the shallowest waters of the coral reefs to the deepest of the Banda Trench (> 7,000 m). The total areas of the Banda Sea for a depth range 0 to >7,000 m was aproximately 591,024.1 km^2. The sampling area of shallow waters in the depth range 0-20 m was 11,484.6 km^2. This depth range was used to assess the reef fish stocks of indicator and major group. For target group reef fish, the depth started at 2 meters. Therefore, the estimated total area for the standing stock of target group at depth of 2-20 m was 7,028.8 km^2.

3.2. Reef fish diversity, abundance and density

Table 1 shows the diversity (numbers of species), the abundance (number of fishes of each species) and the average density (D, in ind.m^{-2} or gram.m^{-2}) for the target, indicator and major groups respectively. These data were originated from both the primary and secondary data from 287 UVC sites. The primary data consisted of quantitative data of all group of reef fishes collected directly in the field mainly from COREMAP (Coral Reef Rehabilitation and Management Project) phase-1 (1998-2004, Initiation stage) [36], COREMAP Phase-2 (2004-2008) [37], and COREMAP CTI (2014-2020) [38]. The secondary data reported mainly only for the target group reef fishes. Since most of this data was qualitative data, thus they were not listed in the table 1. Data in the table 1 was grouped following the 8 regions as shown in the figure 1.

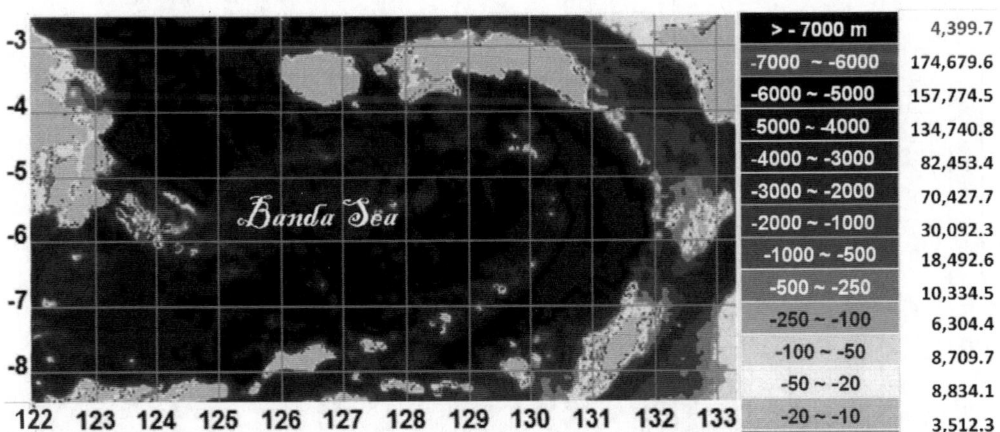

Figure 4. Bathymetry map of the Banda Sea derived from BODC data with the ranges of depth (m) and its areas (km^2) from the deepest of Banda Trench to the shallowest of coral reefs ecosystems.

Table 1. Diversity (number of species), abundance (number of fish in each species) and density (ind.m^{-2}, gram.m^{-2}) of reef fishes in some regions of the Banda Sea.

Location, Year	N	Diversity (Number of Species)				Abundance (Number of Fish)				Average Density				Data Type and Sources
		Target	Indicator	Major	Total	Target	Indicator	Major	Total	Target (ind.m^{-2})	Target (gr.m^{-2})	Indicator (ind.m^{-2})	Major (ind.m^{-2})	
1. North region														
- Lease Islands, 1993	12	57	27	98	182	1,072	390	4,239	5,701	0.858	-	-	-	Primary, (Authors data) [39]
- Ambalau Island, 2008	5	85	28	112	225	7,824	840	9,861	18,525	6.259	1166.2	0.672	7.889	Primary, (Authors data)
- Nusalaut, 2010	6	44	24	97	165	604	188	2,838	3,630	0.403	114.7	0.125	1.284	Primary, (Authors data)
- Ambon, 2011	18	93	26	208	327	3,521	646	9,849	14,016	0.782	149.3	0.144	2.189	Primary (Authors data)
- Pombo Islands, 2014	2	66	18	112	196	557	104	1,985	2,646	0.557	183.5	0.104	3.176	Primary, (Authors data)
Total Stations	31								Average	2.000	403.4	0.261	3.403	
2. Northeast region														
- Watubela Islands, 2007	10	126	29	144	299	14,261	473	16,936	31,670	5.093	726.9	0.169	6.049	Primary (Authors data)
- Watubela Islands, 2017	4	103	21	88	212	8,204	165	5,842	14,117	5.793	870.5	0.118	4.076	Primary, (Authors data)
Total Stations	14								Average	5.443	798.7	0.143	5.061	
3. Mid region														
- Banda Islands, 2012	20	198	32	178	408	22,668	2,276	18,204	43,148	4.534	1,930.0	0.455	7.504	Secondary, [40]
- Banda Islands, 2015	12	98	29	-	127	4,575	843	-	5,418	1.089	191.7	0.201	-	Primary (Authors data)
Total Stations	32								Average	2.811	1060.9	0.328	7.504	
4. East region														
- Kai kecil Islands, 2015	37	-	-	-	-	-	-	-	-	0.736	885.6	-	-	Secondary, [41]
Total Stations	37								Average	0.736	885.6	-	-	
5. South region (Maluku)														
- Southwest Maluku, 2014	20	-	-	-	-	-	-	-	-	2.409	1,030.3	-	-	Secondary, [42]
Total Stations	20								Average	2.409	1,030.3	-	-	
6. South region (NTT)														
- Lembata, 2009	8	52	20	129	201	1,699	319	8,112	10,130	0.850	189.9	0.160	4.056	Primary (Authors data)
- Lamalera, 2013	8	-	27	-	27	-	1,047	-	-	-	-	0.433	-	Secondary, [43]
- East Flores, 2009	10	-	20	-	20	1,230	286	13,795	15,311	0.492	100.2	0.105	5.518	Primary, (Authors data)
- East Flores, 2014	19	-	-	-	-	1,473	-	-	1,473	0.310	54.2	-	-	Secondary, [42]
- Alor, 2014	23	-	-	-	-	3,600	-	-	3,600	0.626	166.7	-	-	Secondary, [42]
Total Stations	73								Average	0.569	127.7	0.232	4.787	
7. Southwest region														
- Wakatobi Islands, 2001	15	-	29	-	30	-	1,318	-	-	-	-	0.239	-	Secondary, [44]
- Wakatobi Islands, 2006	15	-	30	-	30	14,700	1,590	39,210	55,500	3.000	-	0.303	7.469	Secondary, [45]
- Wakatobi Islands, 2007	15	106	32	161	299	7,290	4,125	21,960	33,375	1.389	532.3	1.493	4.193	Secondary, [45]
- Wakatobi Islands, 2015	15	127	28	-	155	8,313	1,535	-	9,848	1.583	290.4	0.279	-	Primary, (Authors data)
- Wakatobi Islands, 2015	15	-	30	-	30	-	1,631	-	1,631	-	-	0.303	-	Secondary, [46]
Total Stations	75								Average	1.486	411.3	0.523	5.826	
8. West region														
- Kendari, 2012	5	30	17	63	110	320	276	4,009	4,513	0.256	79.1	0.123	0.535	Secondary, [47]
Total Stations	5								Average	0.256	79.1	0.123	0.535	
Overall Total Stations	287								Overall Average	1.964	599.6	0.268	4.519	

In terms of reef fish diversity, a total of 630 species consisting of 287 target, 42 indicator, and 301 major species were recorded during UVC survey. The rank of reef fish diversity from the highest to the lowest are the mid-region (Banda Islands), followed by the same rank of northeast region (Watubela Islands) and the southwest regions (Wakatobi Islands), then the north regions (Ambalau Island), the south region (Lembata Island, NTT), and the lowest rank is the east region (islands around Kendari waters), with total species of 408, 299, 299, 225, 201 and 110, respectively (table 1). It seems that the diversity of reef fish in each region correlated with the distance between UVC observation sites to the relatively highly human populated region. For example, the low diversity of reef fish (110 species) was found in the Kendari waters which have a high human population. In contrast, the high diversity of reef fish was found in the region that has low human population such as the Banda Islands, Watubela and Ambalau Islands. The human population number has a linkage with the community activity pressures on coral reef resources.

The diversity of reef fish in the Banda Sea (WPP 714) are about five times higher compared with another location in the Cendrawasih Bay, Papua (WPP 717), such as in the Arui Islands (116 species), south of Biak (105 species), lower Padaido Island (97 species), Upper Upperi Padaido (72 species), and Numfor Island (62 species) [48-50], even perhaps in other WPPs. The richness of reef fish in the Banda Sea is in line with the opinion which states there are about 1,760 species, 3 of which are endemic species [4].

It is somewhat surprising that reef fish from the target fish group in the Banda islands (mid-region of the Banda Sea) have higher biodiversity (198 species) than the major fish group (178 species), because the later ones were often more diverse than target and indicator groups such as in all of other regions in the Banda Sea (table 1) as well as in WPP 717 [48-50]. The reasons are not yet known precisely, but it is likely that overfishing or unfriendly utilizaton of target group fish such as the use of explosives or fish poisoning in this region is not as severe as other regions, either in other regions of Banda Sea or in other Indonesian waters.

In term of abundance, the pattern has almost the same with the reef fish diversity in which the highest abundance was found in the Banda Islands with a total of 43,148, followed by Watubela Islands with a total of 31,670. However, in the other regions, the abundance of reef fishes were less than 20,000, while the lowest was in the islands around the Kendari waters with a total of only 4,513.

With regard to the fish density (D) estimation, it tended to be different than the diversity and abundance patterns. Pointing to the specific sites, the highest average density of the target group reef fish was found in several locations such Ambalau Island (6.3 ind.m^{-2}), followed by the Watubela Islands (5.8 ind.m^{-2}), the Banda Islands (4.5 ind.m^{-2}) and the islands in the southwest of Maluku (2.4 ind.m^{-2}). Lower density ($D < 2.0$ ind.m^{-2}) was found in other locations and the lowest was in the Kendari waters (0.3 ind.m^{-2}). However, the average D for each region was different than sites with the high D of > 2.0 ind.m^{-2} in the Northeast region, the Mid-region, the South regions of Maluku, the North region, while the rest other regions has D of < 1.0 ind.m^{-2}, except the Southwest region with D of 1.5 ind.m^{-2}. The overall average D of target group of reef fishes in the all Banda Sea (WPP 714) was nearly 2.0 ind.m^{-2} (table 2), which is 1.6 higher than in the all Cendrawasih Bay (WPP 717) [50].

Concerning the target group fishes with economic values, Indonesian fisheries statistics authority expressed the stock of commercial fish in the biomass unit (ton) instead of number of individual fish unit. Therefore, the reef fishes belonged to the target group should be expressed in biomass unit as well. The sites with the highest D in biomass unit ($D_{biomass}$) was noted in the Banda Islands, followed by the Ambalau Island, the islands in the southwest Maluku, Kei Kecil Islands, Watubela Islands, and Wakatobi Islands. However, the average $D_{biomass}$ in each region was different compare to specific site. Mid-region and South region of Maluku have higher $D_{biomass}$ of > 1000 gram.m^{-2}; The East and Southwest regions has $D_{biomass}$ of > 500 gram.m^{-2}, while the rest were < 500 gram.m^{-2}, with the west region (Kendari waters) has the lowest $D_{biomass}$ of 79.1 gram.m^{-2}. The overall average $D_{biomass}$ of Banda Sea was 600 gram.m^{-2} (table 2).

The density (D) pattern of the target fish in the Banda Sea differed, both in the individual fish units (D_{fish}) and biomass units ($D_{biomass}$). Reef fishes that have high species diversity and/or high

IOP Conf. Series: Earth and Environmental Science **184** (2018) 012009 doi:10.1088/1755-1315/184/1/012009

abundance or high D_{fish} may not necessarily have high $D_{biomass}$, because the species composition of target fishes and their size in each region varies greatly. For example, fish in Ambalau Island with an abundance only 7,824 has a higher $D_{biomass}$ (1,166.2 grams.m^{-2}) compared to those of high abundance fishes in the Watubela Islands in 2007 (14,621) and 2017 (8,205) and in the Wakatobi Islands in 2015 (8,313). However, the latter two islands waters had a lower $D_{biomass}$ of 726.9, 870.5 and 290.4 gram.m^{-2} respectively (table 1). Thus, it indicated that the reef fishes of the target group in Ambalau Island (north region) were dominated by species with the larger size. A *Naso unicornis* with length of 35 cm and body weight of 904 grams in the Ambalau Island was equal to 11 *Pterocaesio tile* with length of 19 cm and body weight of 85 grams or 6 *Caesio teres* with length of 20 cm and body weight of 154 grams in other regions.

The indicator group fish consists of one family (Chaetodontidae). The Banda and Wakatobi Islands have the highest diversity (33 species), followed by Watubela Islands (29 species), Ambalau Islands (28 species), Lease and Lamarela Islands (27 species), while other regions less than 27 species with the lowest one is in the islands around Kendari waters with only 17 spesies (table 1). In case of the abundance divided by the number of UVC sites, it showed that Wakatobi Islands has the highest average abundance (136 ind.UVC site^{-1}) followed by the Banda Islands (98 ind.UVC site^{-1}), islands around the Kendari waters (55 ind.UVC site^{-1}), and Ambalau-Lease Islands (51 ind.UVC site^{-1}). Some sites were less than 50 ind.UVC^{-1} site such as Wakatobi (46 ind.UVC site^{-1}) and the lowest ones was in the east Flores, NTT (23 ind.UVC site^{-1}). The average D_{fish} of indicator group also showed that the Wakatobi Islands has the highest followed by the Banda Islands, Lease Islands and Watubela Islands with D_{fish} of 0.389, 0.328, 0.261 and 0.143 ind.m^{-2}, respectively, while Islands around Kendari waters as the lowest of 0.123 ind.m^{-2}. The overall average D_{fish} of this group in the Banda Sea was 0.278 ind.m^{-2} (table 2), which was slight higher compare to Cendrawasih Bay, Papua (0.212 ind.m^{-2}).

Table 2. Standing stocks estimation of reef fish in the Banda Sea (WPP 714) for target, indicator and major groups.

Location / UVC survey sites		N	Average Density: Target (ind.m^{-2})	Target (gr.m^{-2})	Indicator (ind.m^{-2})	Major (ind.m^{-2})
1	North region	31	2.000	405.7	0.261	3.403
2	Northeast region	14	5.443	798.7	0.143	5.062
3	Mid region	32	2.811	1,060.9	0.328	7.504
4	East region	37	0.736	885.6	-	-
5	South region (Maluku)	20	2.409	1,030.3	-	-
6	South region (NTT)	73	0.569	127.7	0.232	4.787
7	Southwest region	75	1.486	411.3	0.523	5.826
8	West region	5	0.256	79.1	0.123	0.535
	Total UVC sites	287				
Overall Average Density (D)			1.964	599.6	0.268	4.519
Shallow areas (0-20 m) of the Banda Sea (A)			7,028.8 km$^{2\ a}$		11,484.6 km$^{2\ b}$	
Standing stocks (S) of reef fish			13.805 × 10^{9} individuals	4,214,468 tons	3.078 × 10^{9} individuals	51.899 × 10^{9} individuals

[a] The reef areas for the target group (2-20 m) with the assumption that the fish distribution starts to occur from 2 meters depth.
[b] The reef areas for indicator and major group (0-20 m) with the assumption that the fish distribution starts to occur from the surface layer.

Fishes belong to the indicator group were used as an indicator of coral reefs health. The more diverse and abundance within this group, the more healthy the coral reefs monitored [51]. Many studies show strong correlation between the diversity of indicator fish and the percentages cover of living corals, such as in the Biak and Padaido Islands, Cendrawasih Bay, Papua [52]. Similar results also showed in the southwest coast of Madagascar [51], as well as in Noumea, New Caledonia ([53 and references therein). Therefore, it is no doubt that the fishes classified in the indicator group are very important components to the reef ecosystem.

The diversity of the major group fish in each region of the Banda Sea ranged from 63 (west region, Kendari waters) to 208 species (north region, Ambon Islands) (Table 1). However, the total fish diversity of the major group was 301 species consisted of 26 families, in which Pomacentridae as the most dominant family with 90 species (29.9%) followed by Labridae 75 species (24.9 %), Apogonidae 16 Species (5.3%), and Balistidae 15 species (5%), while the rest of each family within 22 families have species diversity less than 5%. The family Pomacentridae (Damselfish) in the eastern Indonesia waters has the highest species diversity with 123 species, which is the highest in the world [8]. Thus, the Banda Sea represented 73.2% of the world's Pomacentridae.

Meanwhile, the average abundance of major group fish showed that waters of Wakatobi Islands has the highest abundance (2,039 ind.UVC site^{-1}), followed by Watubela Islands (1,629 ind.UVC site^{-1}), south region (NTT) (1,217 ind.UVC site^{-1}), Ambalau-Lease Islands (950 ind.UVC site^{-1}) and suprisingly, the Banda Islands was the lowest (910 ind.UVC site^{-1}). The overall average abundance for the Banda Sea was 1,349 ind.UVC site^{-1}, which was 20, 8 and 12 times higher than the Upper Padaido Islands, Lower Padaido Islands and the south coast of Biak Islands, respectively [50].

The density (D_{fish}) of major groups ranged from 0.535 (west region, Kendari waters) to 7.889 (north region, Ambalau Island). However, the highest average D_{fish} were in the mid-region, southwest and northeast regions with D_{fish} of around 7.0 ind.m^{-2}, while the south region-NTT and the north region have D_{fish} of 3.0 to 4.0 ind.m^{-2} (table 2). The average D_{fish} of major group of the Banda Sea again was 1.4 to 2.0 times higher than that of Cendrawasih Bay waters.

About 70% of fishes belonging to the major group can be classified as ornamental fishes. Despite their generally small size, most of them have economic values. According to the Indonesian Ministry of Marine Affairs and Fisheries (MMAF) data, 70% of ornamental fishes in Singapore market were imported from Indonesia. Furthermore, the data from Indonesia's Statistics Central Agency suggested that Indonesia's ornamental fish export reached US$13 million in 2011 and it is predicted to increase around 30-40% in 2012 [54]. Thus, it is clear that the major group fishes play important roles (as well as target and indicator groups), not only due to their commercial values as ornamental fish but also due to their function in the reef ecosystem. Their high diversity and abundance can also support marine tourism sector. Unfortunately, the COREMAP CTI program has been discontinued to monitor major group fish since 2015 for unknown scientific reasons.

3.3. Reef fish standing stock

The reef fish standing stocks (S) estimation for the Banda Sea (WPP 714) (table 2) was the result of the multiplication of each target, indicator and major groups fish density (D) on each UVC survey site (Table 1) and the areas (A) of shallow water reefs derived from the bathymetry map (figure 4). For target group fish, S was also expressed in fish biomass unit (tons). The S for target, indicator and major groups were 13,805 $\times 10^6$, 3,078 $\times 10^6$ and 51,899 $\times 10^6$ individuals respectively, while the biomass stock of target ones was 4,214,468 tons. This S was rather underestimated especially for the indicator and major group fish, because there is no data available for the west region (Kei Kecil Islands) and the south region Maluku (Southwest Maluku Islands). For target group fish, as explained in the methodology section (cf. section 2.3), we used equations 3 and 4 to estimate TL and used it to convert the number of individual fish to their biomass. The estimated fish TL used here were likely to be shorter than the actual size of the fish during UVC survey. Thus, the target fish stocks were also underestimated from the real ones. Therefore, in terms of reef fish diversity, this study shows that UVC was able to estimate only about 35.8 % out of 1,760 spesies of reef fish in the Banda Sea [4].

In addition, the bias in the standing stock estimation could probably due to the depth where the UVC was carried out (average depth of 5 m), due to equipments constraint. Whereas reef fishes are generally higher in number and larger in size at depths greater than 10 m. The reef fish census cannot always be done at the best of times, when the fish is most visible. Other than that, many nocturnal fishes that can only be seen active at night time as well as planktivorous fish that only appear when the currents carry food particles were less visible during UVC survey [55]. These things show that there are quite a number of limiting factors in the estimation of reef fish stocks by using UVC surveys, which need to be considered for any future monitoring purpose. Nevertheless, UVC survey is a technique that can provide the best available data of the reef fish stocks currently. The availability of these data facilitate the sustainable management of reef fish resources. Moreover, concerning the exploitation management, the underestimated of reef fish stocks can be useful too for conservation purposes.

Table 3 and figure 5 display the list and some photos of reef fishes, which have high standing stocks (S) of the target, indicator, and major groups fish based on the average Density (D) for the whole of Banda Sea. The target group fish in table 3 represents only 12.5% (38 species) of the total diversity target fish (287 species), but account for 61.6% (2,597,052 tons) of the total S (4,214,468 tons). The fish families representing 62.3% S of target group fish starting from the highest S were Acanthuridae (in Indonesian ikan Butana and ikan Kulit pasir) which consisted of *Acanthurus fowleri, A. xanthopterus, A. thampsoni, Naso hexacanthus, N. unicornis* and *N. Lituratus* with total S of 1,004,640 tons, followed by Family of Caesionidae (ikan Pisang-pisang and Ekor Kuning) with *Pterocaesio tile, Caesio caerulaurea, C. teres, C. cuning, Caesio lunaris*, and *Caesio* sp. with total S of 645,645 tons.

Other families with species ranging from 25,000 to <100,000 tons are Lutjanidae (or *ikan Kakap* in Indonesian) consisting of *Lutjanus fulvus, L. bohar, L. gibbus, L. malabaricus, L. argentimaculatus, Macolor macularis* and *M. niger* with estimated total S of 326,097 tons; Family Scaridae (or *ikan Kakatua* in Indonesian) with species *Scarus ghobban, S. dimidiatus, Scarus* sp. and *Chlorurus bleekeri* with total S of 145,829 tons; Family Khyposidae consisted *of Kyphosus cinerascens, K. vaigiensis* and *K. bigibbus* with total S of 102,477 tons; Carangidae with species of *Caranx melampygus* and *Seriola lalandi* have S of 66,457 tons. Other individuals species were *Cheilinus undulatus* (Family Labridae), *Lethrinus miniatus* (Lethrinidae), *Gymnosarda unicolor* (Scomboridae) which is not a reef fish, but a pelagic fish that recorded during UVC, *Mulloidichthys vanicolensis* (Mullidae), *Plectorhinchus chaetodontoides* (Haemulidae) and *Platax teira* (Ephippidae) with S of 82,762; 61,388; 57,180; 50,930; 28,597 and 25,050 tons, respectively. *Cheilinus undulatus*, commonly known as Humphead Wrasse (or *ikan Maming* in Indonesian) is one of the species already listed in The IUCN Red List as Threatened due to its D rarely exceed 20 fish per hectare (0.0020 ind.m^{-2}), but heavily exploited. Even if only moderately exploited, the D of this fish quickly declines to 25% or less of peak densities recorded at no fishing areas [56].

From 11 families listed in table 3, five families (Acanthuridae, Scaridae, Labridae, Khyposidae, and Ephippidae) are herbivorous fishes with a big stock (1,360,758 tons) of more than one third (32.8%) of total S of target group fish, excluding the family of Siganidae (*ikan Baronang* in Indonesian). A ratio pattern between herbivorous and carnivorous fish of 30:70%, indicating a balance in the reef fish population has been reported by the COREMAP-LIPI research team in the Wakatobi Islands, with similar result to this study [46]. Herbivores fishes are known recently to play an important role in maintaining reef ecosystems health. They may help to control and reduce the abundance of certain competitively superior macroalgae from the coral habitat. These superior macroalgae can impede reef resilience if allowed to grow [57,58], particularly when coral reefs were heavily impacted by large scale acute and chronic environmental changes such as El Niño. El Niño caused global coral bleaching in 2016, which makes many reefs in the world collapsed [59] including Indonesian reefs [60]. Thus, a long-term maintenance of reef habitat quality, which underpins fisheries, requires sufficient herbivory stocks [57,58]. However, as many larger species of herbivores fish, such as parrotfish (Scaridae) and surgeonfish (Acanthuridae) are heavily exploited, then it is possible that algae will increase in abundance, start to dominate benthic systems and bring about a decrease in reef biodiversity [59, 61, 62].

IOP Conf. Series: Earth and Environmental Science **184** (2018) 012009 doi:10.1088/1755-1315/184/1/012009

Table 3. List of some fishes with high average Density (D) and Standing stocks (S) of target (tons), Indicator and major (Ind number × 10^6) fish groups in the Banda Sea.

No	Target Group Fish	D	S (tons)	Indicator Group Fish	D	S (ind.)	Major Group Fish	D	S (ind.)
1.	*Acanthurus fowleri* de Beaufort, 1951	50.8	356,899	*Chaetodon kleinii* Bloch,1790	0.034	393	*Pseudanthias huchtii* Bleeker, 1857	0.342	3,386
2.	*Pterocaesio tile* Cuvier, 1830	44.7	314,414	*Hemitaurichthys polylepis* Bleeker,1857	0.031	358	*Chromis ternatensis* Bleeker,1856	0.337	3,333
3.	*Naso hexacanthus* Bleeker, 1855	39.9	280,132	*Heniochus varius* Cuvier,1829	0.015	167	*Pseudanthias squamipinnis* Peters, 1855	0.312	3,091
4.	*Naso thynnoides* Valenciennes, 1835	21.0	147,493	*Chaetodon trifasciatus* Park,1797	0.014	162	*Odonus niger* Ruppell, 1836	0.302	2,985
5.	*Caesio caerulaurea* Lacepede, 1801	15.3	107,826	*Chaetodon trifascialis* Quoy&Gaimard, 1825	0.012	143	*Chromis viridis* Cuvier,1830	0.293	2,897
6.	*Acanthurus xanthopterus* Valenciennes, 1835	14.0	98,133	*Heniochus chrysostomus* Cuvier,1831	0.012	141	*Amblyglyphidodon curacao* Bloch,1787	0.178	1,763
7.	*Caesio teres* Seale, 1906	12.5	87,703	*Chaetodon vagabundus* Linnaeus,1758	0.012	138	*Abudefduf vaigiensis* Quoy & Gaimard,1825	0.163	1,616
8.	*Cheilinus undulatus* Ruppell, 1835	11.8	82,762	*Chaetodon melannotus* Schneider,1801	0.012	134	*Chromis amboinensis* Bleeker,1873	0.112	1,109
9.	*Lutjanus fulvus*Schneider,1801	11.4	79,808	*Chaetodon baronessa* Cuvier,1831	0.011	126	*Melichthys niger* Bloch , 1796	0.104	1,028
10.	*Caesio cuning* Bloch, 1791	9.8	68,570	*Forcipiger longirostris* Jordan & McGregor,1898	0.010	113	*Abudefduf sexfasciatus* Lacepede,1801	0.101	1,002
11.	*Lutjanus bohar* Forsskal, 1775	9.3	65,507	*Forcipiger flavissimus* Jordan & McGregor,1898	0.008	95	*Chromis weberi* Fowler & Bean,1928	0.099	981
12.	*Lutjanus gibbus* Forsskal, 1775	9.3	65,363	*Chaetodon rafflesi* Bennett,1830	0.008	93	*Cirrhilabrus solorensis* Bleeker, 1853	0.088	866
13.	*Lethrinus miniatus* Forster, 1801	8.7	61,388	*Chaetodon punctatofasciatus* Cuvier,1831	0.008	89	*Melichthys vidua* Solander, 1844	0.082	810
14.	*Gymnosarda unicolor* Ruppell, 1836	8.1	57,180	*Chaetodon meyeri* Bloch & Schneider,1801	0.007	79	*Dascyllus reticulatus* Richardson,1846	0.077	760
15.	*Naso unicornis* Forsskal, 1775	7.5	52,601	*Chaetodon lunula* Lacepede,1803	0.006	70	*Chromis margaritifer* Fowler,1946	0.077	757
16.	*Mulloidichthys vanicolensis* Valenciennes,1831	7.2	50,930	*Heniochus singularis* Smith & Radcliffe,1911	0.006	70	*Cirrhilabrus cyanopleura* Bleeker, 1851	0.073	725
17.	*Scarus ghobban* Forsskal, 1775	7.2	50,272	*Chaetodon octofasciatus* Bloch, 1787	0.006	65	*Pomacentrus lepidogenys* Flowler & Bean, 1928	0.071	702
18.	*Acanthurus thampsoni* Fowler, 1938	6.0	42,330	*Chaetodon ocellicaudus* Cuvier,1831	0.006	64	*Pseudanthias tuka* Herre & Montalban, 1927	0.062	618
19.	*Scarus* sp.	5.6	39,206	*Chaetodon unimaculatus* Bloch,1787	0.005	62	*Pseudoanthias dispar* Herre, 1955	0.062	615
20.	*Kyphosus cinerascens* Forsskal,1775	5.5	38,819	*Chaetodon ulietensis* Cuvier,1831	0.005	61	*Cirrhilabrus solorensis* Bleeker, 1853	0.059	582
21.	*Caesio lunaris* Cuvier, 1830	5.5	38,533	*Chaetodon auriga* Forsskal,1775	0.005	59	*Pomacentrus auriventris* Allen, 1991	0.058	575
22.	*Seriola lalandi* Valenciennes, 1833	5.3	36,980	*Chaetodon ephippium* Cuvier,1831	0.005	59	*Pomacentrus moluccensis* Bleeker, 1853	0.058	570
23.	*Kyphosus vaigiensis* Quoy & Gaimard,1825	5.2	36,432	*Chaetodon speculum* Cuvier,1831	0.005	53	*Pseudanthias pleurotaenia* Bleeker, 1857	0.052	514
24.	*Macolor macularis* Fowler, 1931	4.7	32,924	*Chaetodon citrinellus* Cuvier,1831	0.004	51	*Amblyglyphidodon leucogaster* Bleeker,1847	0.051	502
25.	*Chlorurus bleekeri* de Beaufort, 1940	4.4	30,957	*Chaetodon bennetti* Cuvier,1831	0.004	48	*Chromis antripectoralis* Welander & Schultz,1951	0.049	486
26.	*Caranx melampygus* Cuvier, 1833	4.2	29,477	*Chaetodon ornatissimus* Cuvier,1831	0.004	43	*Pomacentrus coelestis* Jordan & Starks, 1901	0.049	485
27.	*Caesio* sp.	4.1	28,599	*Chaetodon lineolatus* Cuvier, 1831	0.003	40	*Chromis xanthura* Bleeker,1854	0.048	470
28.	*Plectorhinchus chaetodontoides* Lacepede, 1801	4.1	28,597	*Heniochus acuminatus* Linnaeus, 1758	0.003	36	*Chromis atripes* Fowler & Bean,1928	0.047	463
29.	*Lutjanus malabaricus* Bloch & Schneider, 1801	4.0	28,439	*Chaetodon adiergastos* Seales,1910	0.003	34	*Dascyllus trimaculatus* Ruppell, 1829	0.045	448
30.	*Lutjanus argentimaculatus* Forsskål, 1775	4.0	28,108	*Chelmon rostratus* Linnaeus, 1758	0.002	25	*Chromis caudalis* Randall,1988	0.045	446
31.	*Kyphosus bigibbus* Lacepède, 1802	3.9	27,226	*Chaetodon oxycephalus* Bleeker,1853	0.0017	19	*Pomacentrus brachialis* Cuvier, 1830	0.042	415
32.	*Naso lituratus* Forster, 1801	3.8	27,051	*Chaetodon mertensii* Cuvier,1831	0.0016	18	*Pomacentrus amboinensis* Bleeker, 1868	0.042	411
33.	*Mocolor niger* Forsskal, 1775	3.7	25,948	*Chaetodon lunulatus* Quoy&Gaimard,1825	0.0015	17	*Amblyglyphidodon aureus* Cuvier,1830	0.040	396
34.	*Scarus dimidiatus* Bleeker,1859	3.6	25,394	*Chaetodon semeion* Bleeker,1855	0.0013	15	*Chrysiptera talboti* Allen,1975	0.032	314
35.	*Platax teira* Forsskal,1775	3.6	25,050	*Heniochus diphreutes* Jordan,1903	0.0010	12	*Chromis analis* Cuvier,1830	0.032	313

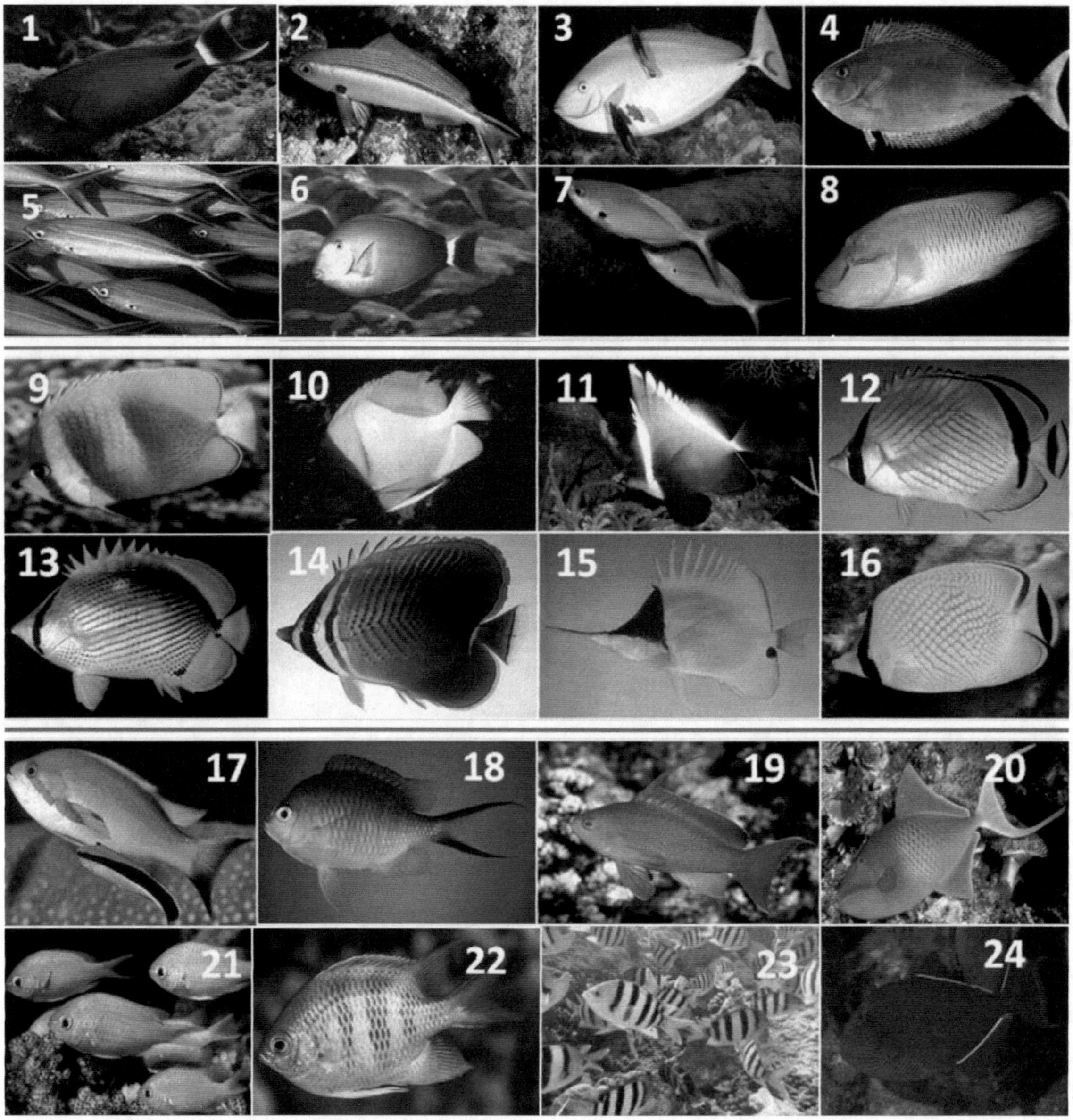

Figure 5. Illustration of predominant target (1-8), indicator (9-16) and major (17-24) fish groups of reef fish stocks of the Banda Sea (see also table 3). Target: 1. *Acanthurus fowleri,* 2. *Pterocaesio tile,* 3. *Naso hexacanthus,* 4. *Naso thynnoides,* 5. *Caesio caerulaurea,* 6. *Acanthurus xanthopterus,* 7. *Caesio teres,* and 8. *Cheilinus undulatus*; Indicator: 9. *Chaetodon kleinii,* 10. *Hemitaurichthys polylepis,* 11. *Heniochus varius,* 12. *Chaetodon trifasciatus,* 13. *Chaetodon trifascialis,* 14. *Chaetodon baronessa,* 15. *Forcipiger longirostris,* and 16. *Chaetodon rafflesi* ; Major : 17. *Pseudanthias huchtii,* 18. *Chromis ternatensis,* 19. *Pseudanthias squamipinnis,* 20. *Odonus niger,* 21. *Chromis viridis,* 22. *Amblyglyphidodon curacao,* 23. *Abudefduf vaigiensis,* and 24. *Melichthys niger*. Photos were made available by Randall J E, Johnson J, Greenfield J and Mayer T and downloaded from [33].

The indicator group fish consisted of one family (Chaetodontidae) dominated by 10 species i.e. *Chaetodon kleinii, Hemitaurichthys polylepis, Heniochus varius, C. trifasciatus, C. trifascialis Heniochus chrysostomus, C. vagabundus, C. melannotus* and *C. baronessa* which has a standing stock (S) > 100 $\times 10^6$ ind. (table 3; figure 5). Thus, the total S only for these 10 species was 1,872 $\times 10^6$ ind.

or composed 60.8% of the total indicator group fish. For stock ranged from 50 to100 × 10^6 ind. consisted of 14 species with the total stock 919 ×10^6 (29.8%). While for the stock less than 50 × 10^6 ind. consisted of 18 species with the total of S 346 × 10^6 ind. (11.2%), including six species that were not listed in table 3 i.e. *Coradion chrysozonus, C. xanthurus, Parachaetodon ocellatus, C. melanopus, C. pelewensis* and lowest *C. altivelis.*

C. cleinii has high abundance and D not only in the Banda Sea but also widely distributed in all Indonesian waters [36, 63-65]. It seems that this fish is a cosmopolitan fish [43, 47]. In addition, the presence and high density or abundance of *C. kleinii* as shown in table 3 can be considered as the indicator for a good condition of coral reef [43]. On the contrary, the high abundance of *C. octofasciatus* indicates that the coral reefs in a degraded condition because of low water transparency, sedimentation or low salinity [43]. Table 4 showed clearly the evidence that waters with relatively good reef conditions such as in the Lamalera, Lembata, East Flores, Wakatobi, Banda, Lease and Watubela Islands had a high abundance of *C kleinii,* but low on *S. octofasciatus.* The opposite condition was found in the Kendari waters with coral conditions that were not as good as those mentioned above with the low abundance of *C. kleinii*, but have a very high abundance of *C. octofasciatus* (about 32 times higher).

Furthermore, as an indicator of coral health, the bright and beautiful colors of family Chaetodontidae make them also classified as ornamental fish, like the major fish group. Some species that are often used or sold as ornamental fishes are *Chelmon rostratus, Heniochus diphreutes, Forcipiger flavissimus, F. longirostris, Chaetodon lunula, C. punctatofasciatus, C. mertensii, C. unimaculatus, C. rafflesii, C. ulietensis, C. falcula, C. auriga, C. Rafflesi dan C. Semilarvatus* [67]. However, most of these species have low S, except *F. Longirostris* and *F. Flavissimus* with S of 113.1 and 95.2 × 10^6 ind., respectively (table 3).

The major fish group (mostly ornamental fish) consisted of 26 families with the total S of 50,027 × 10^6 ind. However, only four families dominated in this group i.e Serranidae, Pomacentridae, Balistidae, and Labridae. Species of these families with S ranging from 1,000 to > 3,000 × 10^6 ind. were *Pseudanthias huchtii, Chromis ternatensis, P. Squamipinnis, Odonus niger, C. viridis, Amblyglyphidodon curacao, Abudefduf vaigiensis, C. amboinensis, Melichthys niger* and *Abudefduf sexfasciatus* (table 3) with the total of S 22,210 × 10^6 ind., composing 44.4 % of the total major fish group. Major fish with S ranged from 500 to < 1,000 × 10^6 ind., 250 to < 500 × 10^6 ind., and < 250 × 10^6 ind. (not listed in table 3) were 9,576 × 10^6 ind. (19.1%), 5,499 × 10^6 ind. (11 0%), and 12,757 × 10^6 ind. (25.5%) respectively. From table 3, 19 species of Pomacentridae dominated the stocks followed by species of Serranidae, three species of Balistidae and five species of Labridae with the total S of each family were 21,474 × 10^6 ind. (42.9%), 8,225 × 10^6 ind. (16.4 %), 4,824 ind. (9.6%) and 2,764 ind. (5.5%), while the rest composed of 230 species belonging to 26 families with the total S of 12,757 ind. (25.5%).

Table 4. Density and abundance of *C. kleinii* and *C. octofascitus* as indicators of coral reefs condition.

Region (locations) in the Banda Sea	*C. kleinii*		*C. octofasciatus*	
	D (ind.m^{-2})	Abundance (ind.)	D (ind.m^{-2})	Abundance (ind.)
North Region (Ambalau-Lease Islands)	0.025	287,115,000	0.001	11,484,600
Northeast Region (Watubela Islands)	0.019	218,207,400	0.000	-
Mid region (Banda Islands)	0.028	321,568,800	0.000	-
South Region, NTT (Lamalera, Lembata, east Flores)	0.060	689,076,000	0.001	11,484,600
Southwest region (Wakatobi Islands)	0.059	677,591,400	0.000	-
West Region (Kendari waters)	0.015	172,269,000	0.032	367,507,200

The major fish group is likely to have much higher stocks, but available data are limited. There was no available data of the major fish in the east region (Kei Kecil Islands) and south region of Maluku (table 2) except for the total number of fish belonging to this group that already calculated from each family, such as in the west region or Islands in the Kendari waters [49]. In other regions, only qualitative data were available for major group fishes by showing the presence of each species and the total number of this group [47]. It appears that the major fish group data was more negligible than the target and indicator ones. This drawback may be one of the main reasons to stop the major fish census on the CTI COREMAP program. Thus, until now the role of this group of fish on the health of coral reefs is still not known clearly.

Further, the average density (D) of reef fish in the Banda Sea was compared with the density in the Cendrawasih Bay, Papua (WPP 717) which has similar geographical conditions of many small islands surrounded by deep seas such as Numfor, Biak, Padaido Islands (Biak-Numfor District) and the Arui Islands (Supiori district). The average density of reef fish in the Banda Sea was about 1.62, 1.29 and 1.88 times higher for the target, indicator and major fish group respectively than those in the Cendrawasih Bay (appendix 1). This was due probably to the influences of seasonal upwelling phenomenon monitored using long-term (2003-2017) multi-temporal of average seasonal data of Chlorophyll-a concentration (Chl-a) and sea surface temperature (SST).

Satellite image of Aqua-MODIS was used to describe this phenomenon (figure 6). During east monsoon (June-July-August, JJA), Chl-a ranged from 0.2 to 1.5 $\mu g.l^{-1}$ or 2-6 times higher which indicated a high productivity of the Banda Sea (figure 6) than to Chl-a on west monsoon (December-January-February, DJF) and transitional-I season (March-April-May, MAM). The distribution of sea surface temperature (SST) in JJA was dominated by low SST (< 28.0°C), indicating an upwelling, but very high during DJF and MAM seasons where SST ranged from 30 to more than 31.5°C. During transitional-II season (September-October-November, SON), upwelling starts to be low indicated by a decrease of Chl-a (0.2-0.5 $\mu g.l^{-1}$), while SST starts to increase (29-30°C).

This upwelling phenomenon during JJA season was due to a regional wind which drives a motion associated with the monsoon climate [66]. This indicator could be observed by Aqua-MODIS satellite as similar with this study [67]. Upwelling event was also well studied by using the indicators of low sea temperatures and dissolved oxygen (DO), but high in salinity, nutrients and Chl-a concentration at near surface layer [68] followed by high phytoplankton and zooplankton abundance [69-71]. The abundance of zooplankton was higher just after upwelling season or during SON [72]. Once phytoplankton and zooplankton are abundant, it will support the reef fishes of the Banda Sea with many foods in the food chain, as well as the formation of fishing ground for larger fish, i.e Skipjack and Yellowfin Tuna [73].

The results of this study indicated that reef fish stocks in the Banda Sea (WPP 714) were high, so it can support the livelihood of small island communities around the Banda Sea. However, sustainable management of reef fish stocks is required to preserve them. Thus, the communities must exploit the reef fish resources in a more sustainable way. Unsustainable fishing practices such as the commonly used explosives, poisons or bad fishing gears in many small islands all around Indonesian waters (such as in the Padaido Islands, Papua, WPP 717) [50] will make the stocks of reef fishes lower than in the Banda Sea (WPP 714) and thus must be avoided. In recent decades, the spearfishing activities using a harpoon and scuba gear, particularly at night in many reefs in the world tend to increase. These activities are focused more on herbivorous (i.e. Acanthuridae, Labridae, and Scaride) fish catches [74], especially the big size ones. This is alarming to the coral ecosystem scientists because it leads to not only the reduction of the reef fish stocks but also to the declining health of coral reefs [55,75].

Looking back at table 1, there was a tendency of reef fishes stock depletion over time. For example, in the north region of the Banda Sea, the average density (D) of the target group fish of Lease Islands (Ambon, Haruku, Saparua and Nusa Laut Islands), decreased from 0.858 $ind.m^{-2}$ in 1993 to 0.403 (Nusalaut Island, 2010), 0.782 (Ambon Island, 2011) and 0.577 $ind.m^{-2}$ (Pombo Island, 2017). In the northeast region, the average density of indicator and major fish groups of the Watubela islands decreased from 0.169 to 0.118 and from 6.049 to 4,070 $ind.m^{-2}$ during 2007 to 2017 respectively. The same pattern was found for the target and indicator fish groups of the Banda Islands where it decreased from 4.534 to 1.089 ind.m^{-2} and from 0.455 to 0.201 ind.m^{-2} during 2007 to 2017. While in the south region of the east Flores Islands (NTT), the target fish group decreased from 0.492 to 0.310 ind.m^{-2} from 2009 to 2014. These tendencies were similar with the reef fish stock condition in the most of Biak, Numfor and Padaido Islands, and Arui Islands of Cendrawasih Bay, Papua (WPP 717) [50]. Therefore, the management of fish resources needs to be done seriously where one of the applicable methods is the establishment of a marine protected area (DPL).

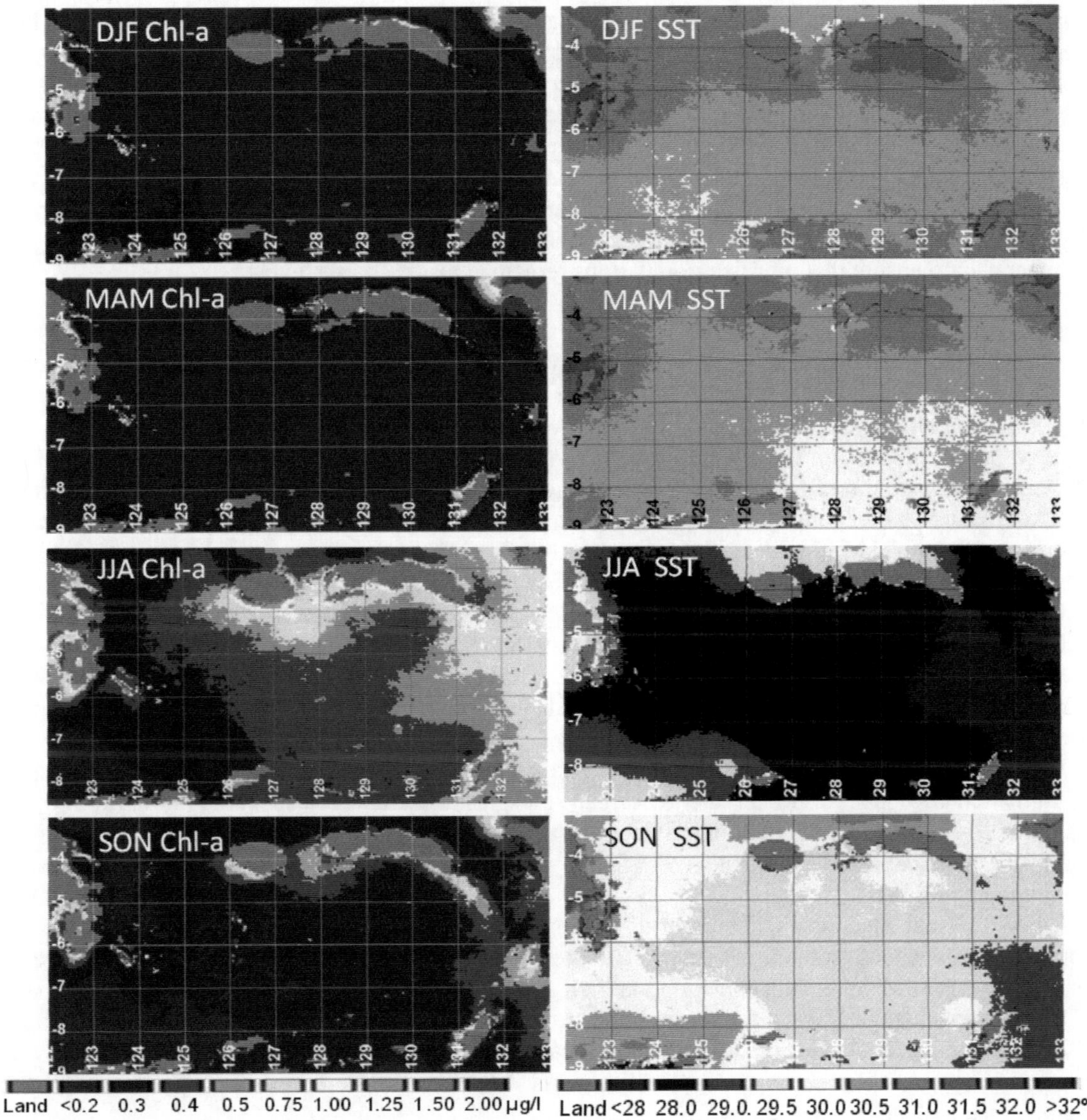

Figure 6. Mean seasonal chlorophyll-a consentration (μg.l^{-1}) (left) and SST (°C) (right) in the Banda Sea acquired from Aqua-MODIS Satellite, 2003-2017.

Results of the study on the effectiveness of three types of DPL in Indonesia and Papua New Guinea (PNG), namely: National Park (TN), co-managed DPL and traditional DPL showed that the traditional DPL type in Muluk, Ahus (PNG) and Kakalotan Island (North Sulawesi, Indonesia) were the most effective in conserving the reef fish resources. This was indicated by higher biomass and fish size of target group fish inside DPL than outside DPL [76]. Those DPLs were small in size (33-58 ha), managed by local communities and periodically closed following local wisdom rules (e.g Sasi in Maluku and Papua). Similar results were found on 20 DPLs established by Coremap LIPI in 2008 in the south coast of Biak Island (4 DPLs), lower Padaido (8 DPLs) and upper Padaido islands (8 DPLs), which were traditional DPL [65] with small areas of 15-20 ha (length ± 500m, width ± 400 m). It seemed that these DPLs were more effective in conserving the reef fishes. The average density (D) of target, indicator and major groups fish within these DPLs were 2.8-3.6, 1.5-3.0 and 1.8 - 4.4 times higher than the fish outside the DPLs respectively [77].

In the Kei Kecil Islands, several NGOs, local governments and universities are working together to set up a co-managed DPL called "Small Island Park". They monitor the condition of fish in the DPL. The DPL consists of 3 zones: 1) utilization zones, 2) no-take zones and 3) outer zones. Results showed that herbivorous (Acanthuridae, Scaridae, and Siganidae) and some carnivorous (Serranidae, Lethrinidae and Nemipteridae) fishes in the no-take zone area have the highest fish density (ind.ha^{-1}) and biomass (kg.ha^{-1}), although the overall of total fish were more abundant in the utilization zone [41]. Therefore, regarding the previously cited references [76,77], it is clear that the establishment of DPLs, particularly no-take zones in many locations around small islands in the Banda Sea is highly recommended, to ensure that reef fish populations remain at a sustainable level.

4. Conclusion

The estimation of reef fish stocks for Indonesian waters is still very rare due to a highly fragmented data. The data is almost not recorded in Indonesian fishery statistics, except for a certain species only. In this study we have conducted stock assessments of reef fishes for the most islands around the Banda Sea representing Indonesian fisheries management region of WPP 714. The advantage of this study is that we have estimated reef fish stocks to the species level, where this has not been done before for a wide area on a WPP region level. Given the availability of this data, it will support the fisheries decision makers in managing reef fish resources, such as the stock of a species or group of fish (yellowtail/banana fusilier, grouper, snapper and other fishes) in order to be exploited sustainably.

The small islands in the Banda Sea have a high diversity of reef fishes.There were at least 630 species comprised of 287, 42 and 301 species of the target, indicator and major groups, respectively including the estimated stock data of each species. Moreover, the fish stocks of indicators and major groups that also have potentials as ornamental fish to support marine tourism with its own market share.

The stock assessment of reef fish of Banda Sea was conducted using UVC survey. The results of this study tends to underestimate the reef fish stocks and there are also some limitations to be considered. Nevertheless, this method has at least provided the best reef fish stocks data that can be used for the sustainable management of reef fish resources. In the near future, the use of this method is needed to be improved.

All the reef fish resources will support the livelihoods and the welfare of the local community. As a consequence, the local community should use the reef fish resources in a more sustainable and eco-friendly way. Furthermore, in order to make the fish stocks to be available and sustainable, many DPLs need to be established. The co-management of DPLs as has been done in the Kei Kecil Islands and/or traditional DPLs that are supported by local people (traditional one) has proven effective in conserving fish resources. Thus, DPL establishments should be constantly proposed and developed to the entire small islands in the Banda Sea.

Considering the vast area of Indonesian waters (11 fisheries management area, WPP), this study has initiated fish stocks assessment in the Banda Sea (WPP 714) using the best available data. Some data were also available for the Cendrawasih Bay (WPP 717) although limited. There are still nine WPPs that are still unknown. Therefore, the stock assessment of reef fish in other WPPs must be started immediately by using available secondary data. Thus, the estimation data of reef fish stocks abundance will be a needed important addition to the Indonesian fisheries statistics comprising only the pelagic and demersal fish stocks so far. The existence of reef fish stocks data throughout Indonesia then can be used for the sustainable reef fish management purposes.

Acknowledgements
We would like to thank Yance Hehuat and La Pay who were tirelessly and always ready to collect and provide a lot of valuable reef fish data as they searched from many sources, both primary and secondary data. We also express our gratitude to the British Oceanographic Data Center (BODC) that made the bathymetry data available from the General Bathymetric Chart of the Ocean (GEBCO). The analysis and visualizations used in this study (Figure 6) were produced with Giovanni's online data system, developed and managed by DISC GES NASA. Last but not least, we extend our thanks to Drs/Ms/Mrs J E Randall, J Johnson, M McVicar, and T Mayer, who provide excellent reef fish photos and allow us to download from [33]. Lastly, we would like to thank the three reviewers who have taken the time to read this manuscript carefully and provide valuable suggestions that make this paper much better.

Appendix 1. Standing stocks of reef fish in the Cendrawasih Bay (WPP 717) for target, indicator and major groups as a comparison to the Banda Sea (WPP 714).

Location (UVC sites).Year^{-1}	N	Target (ind.m^{-2})	Indicator (ind.m^{-2})	Major (ind.m^{-2})	Ref.
Biak-Numfor District					
- Numfor Island 2011-2012	31	0.71	0.07	0.57	[78]
Average		0.71	0.07	0.57	
- Biak Island, 1995	4	1.53	0.49	6.37	[48]
- Biak Island 2008, No-Take Zone (NTZ)	4	1.78	0.79	2.23	[79]
- Biak Island, 2010-2012	21	0.58	0.17	1.06	[50]
Average		1.30	0.48	3.22	
- Lower Padaido Islands, 2001	6	1.19	0.09	2.34	[49]
- Lower Padaido Islands, 2010-2012	25	0.32	0.09	1.25	[61]
- Lower Padaido Islands, 2008, NTZ Area	8	1.27	0.40	4.35	[79]
Average		0.93	0.19	2.65	
- Upper Padaido Islands, 2001	6	0.79	0.11	4.44	[49]
- Upper Padaido Islands, 2010-2012	31	0.23	0.05	0.70	[50]
- Upper Padaido Islands, 2008, NTZ Area	8	0.67	0.17	2.12	[79]
Average		0.56	0.11	2.42	
Supiori District					
- Aruri Islands, 1995	6	1.22	0.16	1.96	[48]
- Aruri Islands, 2009	20	3.56	0.29	4.16	[80]
- Aruri Islands, 2013	25	2.70	0.17	2.96	[81]
Average		2.49	0.21	3.03	
Total UVC survey sites	209	-	-	-	
Overall Average Density		1.212	0.212	2.378	

References

[1] The Editors of Encyclopaedia Britannica 2017 *Banda Sea* [Internet] accessed Oct 2017 from http://www.britannica.com/place/Banda-Sea
[2] Hutomo M and Moosa M.K 2005 Indonesian Marine and coastal biodiversity: Present Status *Indian J. Mar. Sci.* **34**(1) 88-97
[3] Spalding M D *et al.* 2007 Marine Ecoregions of the World: A Bioregionalization of Coastal and Shelf Areas *BioScience* **57**(7) 573-83
[4] Huffard C L, Erdmann M V and Gunawan T R P (eds) 2012 *Geographic priorities for marine biodiversity conservation in Indonesia* (Jakarta: Ministry of Marine Affairs and Fisheries and Marine Protected Areas Governance Program) 105 pp

[5] Green A L and Mous P J 2008 Delineating the Coral Triangle, its Ecoregions and Functional Seascapes. Version 5.0. *TNC Coral Triangle Prog. Rep.* 1/08 p 44
[6] CT Atlas 2017 *About Coral Triangle* [Internet] accessed Oct 2017 from http://ctatlas.reefbase.org/coraltriangle.aspx
[7] TNC 2008 *Coral Triangle facts, figures and Calculations, Part II: Patterns of Biodiversity and Endemism* [Internet] accessed Oct 2017 from ctatlas.reefbase.org/pdf/part%20II%20%20 biodiversity%20stats.pdf p 6
[8] Allen G R and Adrim M 2003 Coral Reef Fishes of Indonesia *Zoological Studies* **42**(1) 1-72
[9] Burke L, Selig E and Spalding M (eds) 2012 *Reef at risk in Southeast Asia* (Washington DC: World Resources Institute) 72 pp
[10] Suharsono 2007 *Pengelolan terumbu karang di Indonesia* Research Professor Confirmation Lecture in Marine Biology (Jakarta: RCO LIPI) p 112
[11] Monk K A, De Fretes Y and Liley G H 2000 *Ekologi Nusa Tenggara dan Maluku* ed Kartikasari (Jakarta: Prenhalindo) p 966
[12] Dudley R G and Harris K C 1987 The fisheries statistics system of Java, Indonesia: Operational realities in a developing country *Aquacult. Res.* **18**(4) 365-74
[13] Wouthuyzen S and Lorwens J 2015 Pendugaan Stok ikan Karang *Protokol Pengkajian Stok Sumber Daya Ikan* eds A Ghofar, P Martosubroto and Wudianto (Jakarta: KomNasKan PSDI KKP) pp 125-66
[14] Vanderstraete T, Goossens R and Ghabour T K 2003 Remote sensing as a tool for bathymetric mapping of coral reefs in the Red Sea (Hurghada – Egypt) *Belgeo* **3** 257-68
[15] Lampe M, Demmalino E B, Neil M and Jompa J 2017 Main drives and alternative solution for destructive fishing in South Sulawesi-Indonesia: Lesson learned from Spermonde Archipelago, Taka Bonerate, and Sembilan Island, *Sci.Internat.(Lahore)* **29**(1) 159-64
[16] Wahid A 2015 Social impact of blast fishing and Cyanide fishing to fishermen *Internat. Conf. Sci. Technol. Humanity* p 88-98
[17] Wilson S K, Fisher R, Pratchewtt M S, Graham N A J, Dulvy N K, Turner R A, Cakacaka A, Pollunin N V C and Rushton S P 2008 Exploitation and habitat degradation as agents of change within coral reef fish communities *Glob. Change Biol.* **14** 2796–809
[18] Gulland J A 1983 *Fish stock assessment: A manual of a basic method* (Cambridge: John Wiley & Sons) pp 223
[19] Sparre P and Venema S C 1998 Introduction to tropical fish stock assessment Part 1 *Manual FAO Fisheries Technical Paper* No 306.1 Rev. 2 (Rome: FAO) p 407
[20] NOAA Fisheries 2012 *Fish Stock Assessment 101 Series: Part 1—Data Required forAssessing U.S.Fish Stocks* [Internet] accessed Oct 2017 from: http://www.nmfs.noaa.gov/stories/2012/05/05_23_12stock_assessment_101_part1 .html
[21] British Oceanographic Data Centre (BODC) 2017 *General Bathymetric Chart of the Oceans* (GEBCO) [Internet] accessed Oct 2017 from https://www.gebco.net/
[22] Dartnall AJ and Jones MS 1986 *A manual of survey methods for living resources in coastal areas* (Townsville: AIMS & ASEAN & Australian Development Assistance Bureau) 80 pp
[23] Harford A R and Thomson A A 1994 *Visual Census of survey of Reef Fish Standard Operational Procedure No.3* (Townsville: AIMS) p 22
[24] Allen G R 1991 *Damselfishes of the World* (Germany: Mergus Press) 271 p
[25] Myers R F 1989 *Micronesian Reef Fishes* Coral Graphics Agana Guam
[26] Randall J E, Allen G R and Steene R C 1990 *Fishes of the Great Barrier Reef* (Bathurst: Crawford House, Australia)
[27] Kuiter R H 1993 The complete diver's and fisherman's guide to coastal fishes of the south-eastern (Bathurst: Crawford House, Australia) p 437
[28] Lieske E and Myers R 1995 *Coral reef Fishes, Indo-Pacific&Carribean* Collins Pocket Guide p 400

[29] English S Wilkinson C and Baker V 1994 *Survey Manual for Tropical Marine Resources* Asean–Australia Marine Science Project: Living Coastal Resource (Townsville: Australian Institute of Marine Science) p 368

[30] Kulbicki M, Tham G M, Thollot P and Wanitiez L 1993 *Length-weight relationships of fish from lagoon of New Caledonia* (Naga: ICLARM) pp 26-30

[31] Kulbicki M, Guillemot N and Amand M 2005 A general approach to length-weight relationships for New Caledonian fishes *Cybium* **29**(3) 235-52

[32] Kulbicki M, Bozec Y M and Green A 2005 Implication of biogeography in the use of butterflyfishes (Chaetodontidae) as indicator for Western and Central Pasific Area *Aquat. Conserv. Mar. Freshw. Ecosyst.* **15** S109-26

[33] Froese R and Pauly D (eds) 2018 FishBase [Internet] Accessed on July 2018 from www.fishbase.org version (06/2018)

[34] Bohnsack J A and Harper D E 1988 Length-weight relationships of selected marine reef fishes from the southeastern United States and the Caribbean *NOAA Technic. Memor. NMFSSEFC-215* p 31

[35] Yanagawa H 1994 *Length-weight relationship of Gulf of Thailand fishes.* Fishbyte Section (Naga: ICLARM) p 48-52

[36] COREMAP 2017 *COREMAP PHASE I* [Internet] Accessed on Nov 2017 from http://coremap.or.id/berita/15

[37] COREMAP 2017 *COREMAP PHASE II* [Internet] Accessed on Nov 2017 from http://coremap.or.id/berita/16

[38] COREMAP-CTI 2017 *Coral Reef Rehabilitation and Management Program-Coral Triangle Initiative (COREMAP-CTI)* [Internet] Accessed on Nov 2017 from http://projects.worldbank. org/ P127813/coral-reef-rehabilitation-management-program-phase-iii?lang=en

[39] La Tanda and Hukom F D 1994 Studi pendahuluan komunitas karang di beberapa perairan terumbu karang Kepulauan Lease Pelatihan (lanjutan) *Metodologi Penelitian Penentuan Kondisi Terumbu Karang* (Jakarta: Pusat Penelitian Oseanografi LIPI) p 16

[40] Muljadi A H and Hehuat Y 2012 Status of Commercial Fish Populations in the Banda Islands *Marine Rapid Assessment of the Banda Islands, Maluku Tengah, Indonesia* (Bali:Coral Triangle Center) chapter 7 pp 98-158

[41] Prabuning D, Putra M I H, Amkieltiela, Abdillah T, Estradivari and Harris J 2016 *Pengamatan Komunitas Bentik dan Ikan Karang di Kawasan Konservasi Perairan dan Pulau-Pulau Kei Kecil, Pulau-Pulau, dan Perairan Sekitarnya-Maluku Tenggara* (Tual: WWF Indonesia-Kei Office) p 27

[42] Setiawan F *et al.* 2017 Reef Fishes Biodiversity of Lesser Sunda-Banda Seascape (East Flores, Alor and Southwest Maluku District) *Jurnal Kelautan* **10**(1) 1-20

[43] Wibowo K, Adrim M, and Makatipu P C 2013 Comunity Structure of *Chaetodontidae* in the west of Banda Sea *Mar. Res. Indonesia* **38**(1)1–8

[44] Badrudin, Suharti S R, Yahmantoro, Suprihanto I 2003 Indeks keanekaragaman hayati ikan Kepe-kepe *(Chaetodontidae)* di perairan Wakatobi, Sulawesi Tenggara *JPPI Edisi Sumberdaya dan Penangkapan* **9**(7) 67-73

[45] CRITIC COREMAP II LIPI Team 2007 *Monitoring kesehatan terumbu karang Kabupaten Wakatobi tahun 2007* (Jakarta: COREMAP II LIPI) 70 pp

[46] Tuti Y *et al.* 2015 *Monitoring Terumbu Karang dan Ekosistem terkait di Kabupaten Wakatobi* (Jakarta: COREMAP-CTI) 90 pp

[47] Adrim M, Harahap S A, and Wibowo K 2012 Struktur Komunitas Ikan Karang di Perairan Kendari *Ilmu Kelautan* **17**(3) 154-163

[48] Wouthuyzen S 1995 Status ekosistim wilayah pesisir Pulau Biak dan sekitarnya *Laporan Akhir 1994/1995* (Ambon: Balai Penelitian dan Pengembangan Sumber daya Laut Ambon) 138 pp

[49] Wouthuyzen S, Sapulete D, Peristiwadi T, Hukom F, La Tanda, Papalia, S and Lorwens J 2001 Pengkajian metodologi pendugaan stok ikan karang di P. Biak dan P.P. Padaido *Laporan Akhir 2001* (Ambon: Balai Penelitian dan Pengembangan Sumber daya Laut Ambon) 129 pp
[50] Lorwens J, La Tanda and Wouthuyzen S 2013 Pendugaan stok ikan karang di Perairan Biak dan Kepulauan Padaido, Papua *Oseanologi dan Limnologi di Indonesia* **39**(5) 347-68
[51] Wu A, Parvathaneni A and Polcyn M 2015 Biome Project: Open Ocean vs. Deep Ocean [Internet] accesed on Nov 2017 from https://frontier.ac.uk/publications/files/2010_10_14_17_00_53_004.pdf
[52] Lorwens J 2011 Hubungan antara ikan indikator *(Chaetodontidae*) dan kondisi karang di pesisir Pulau Biak dan Kepulauan Padaido *Jurnal Literatur Perikanan Indonesia* **17**(2)147-56
[53] Bozec Y M, Dole´Dec S and Kulbicki M 2005 An analysis of fish-habitat associations on disturb Coral reefs: chaetodontid fishes in New Caledonia *J. Fish Biol.* **66** 966–82
[54] Indonesia Tropical Fish Exporter 2017 *IndoFish Exporter 2017* [Internet] accessed on November 2017 from https://indofishexporter.com/about-us/indonesia/
[55] Laboute P A 2013 *Rapid Marine Biodiversity Assessment of the Northeastern Lagoon from Touho to Ponérihouen, Province Nord, New Caledonia* (New Caledonia: Conservation International) p 161-68
[56] Russell B 2004 C*heilinus undulatus. The IUCN Red List of Threatened Species 2004:e.T4592A11023949* (Grouper & Wrasse Specialist Group) [Internet] accesed Nov 2017 from http://dx.doi.org/10.2305/ IUCN.UK.2004.RLTS. T4592A110 23949.en
[57] Ogden JC and Lobel P S 1978 The role of herbivorous fishes and urchins in coral reef communities *Env. Biol. Fish.* **3**(1) 49-63
[58] Mumby P J 2016 Stratifying herbivore fisheries by habitat to avoid ecosystem overfishing of coral reefs. *Fish Fish.* **17** 266-278. doi:10.1111/faf.12078
[59] Edwards C B, Friedlander A M, Green A G, Hardt M J, Sala E, Sweatman H P, Williams I D, Zgliczynski B, Sandin S E, and Smith J E 2013 Global assessment of the status of coral reef herbivorous fishes: evidence for fishing effects. *Proc. R. Soc.* B 20131835 http://dx.doi.org/10.1098/rspb.2013.1835
[60] Wouthuyzen S, Abrar M and Lorwens J 2018 A comparison between the 2010 and 2016 El-Ninō induced coral bleaching in the Indonesian waters *IOP Conf. Ser.: Earth Environ. Sci.* 118 012051 16 pp
[61] Plass-Johnson J G, Ferse S C A, Jompa J, Wild C and Teichberg M 2015 Fish herbivory as key ecological function in a heavily degraded coral reef system *Limnol. Oceanog.* **60**(4)1382-91
[62] Dell C L A, Longo G O and Hay M E 2016 Positive Feedbacks Enhance Macroalgal Resilience on Degraded Coral Reefs *PLoS ONE* **11**(5):e0155049
[63] Adrim, M 1999 Ikan-ikan suku *Chaetodontidae, Pomacanthidae, Pomacentridae*, yang ada dalam koleksi refrensi Pusat Penelitian Oseanografi – LIPI ed K M Moosa *Katalog koleksi biota laut Jilid-1 (Ikan, Koralia dan Krustacea)* (Jakarta: Pusat Penelitian Oseanografi – LIPI) p 16
[64] Adrim M 2007 The reef fish community in Enggano Islands, Bengkulu Districts *Oseanologi dan Limnologi di Indonesia* **38** 139-58
[65] Hukom F D 2016 Sumberdaya ikan hias perairan karang Maluku Utara *Pengkajian Stok dan Potensi Ikan Karang di Wilayah Propinsi Maluku Utara* (Ternate: Maluku Utara Provincial Bureau for Development Planning [BAPPEDA]) p 265
[66] Ratnawati H I, Hidayat R, Bey A and June T 2016 Upwelling in the Banda Sea and the south coast of Java and its relationship with ENSO dan IOD *Omni-Akuatika* 12(3) 119-30
[67] Sukresno B and Suniada K I 2008 Observasi pengaruh ENSO terhadap produktifitas primer dan potensi perikanan dengan menggunakan data satelit di Laut Banda *Majalah Ilmiah Globe* **10**(2) 97-107
[68] Wouthuyzen S 2002 Study on the upwelling in the Seram and Banda Seas *Oseanologi dan Limnologi di Indonesia* **34** 17-35

[69] Sediadi A 2004 The Effect of Upwelling on Distribution and Abundance of Phytoplankton in Indonesia *Makara Sains* **8**(2) 43-51

[70] Yusuf S A and Wouthuyzen S 1997 Zooplankton Abundance in the Seram and Banda Seas *Proc. Marine Seminar of LIPI-UNHAS*, Ambon, 4-6 July 1997 p 218-26

[71] Tapilatu Y H 2016 Profil Oseanografi Biologi Laut Banda: Sebuah Tinjauan Kritis *Omni Akuatika* **12**(2) 58–66

[72] Baars M A, Sutomo A B, Oosterhuis S S and Arinardi O H 1990 Zooplankton abundance in the eastern Banda Sea and northern Arafura Sea during and after the upwelling season, August 1984-February 1985 *Neth. J. Sea Res.* **25**(4) 527-43

[73] Wouthuyzen S, Hutahaean W and Nanlohy A 1998 Studi tentang penangkapan ikan cakalang, *Katsuwonus pelamis* di perairan Maluku Tengah *Prosiding Seminar Nasional Kelautan KTI ke-II* (Makassar: UNHAS) p 7

[74] Chavarro S B, Mumby P J and Golbuu Y 2014 Changes in the spear fishery of herbivores associated with closed grouper season in Palau, Micronesia *Anim. Conservat.* **17** 133–43

[75] Lindfield S J, McIlwain J L, Harvey E S 2014 Depth Refuge and the Impacts of SCUBA Spearfishing on Coral Reef Fishes *PLoS ONE* **9**(3): e92628

[78] McClanahan T R, Marnane M J, Cinner J E and Kiene W E 2006 A Comparison of Marine Protected Areas and alternative approach to coral-reef management *Curr. Biol.* **16**(4) p 1408-13

[77] Wouthuyzen S, Lorwens J and Hukom F D 2016 Is fish sanctuary effective to conserve reef fishes? A case study in the Biak-Numfor and Supiori Districts, Papua *Jurnal Penelitian Perikanan Indonesia* (in Indonesian) **22**(4) 271-84

[78] Lorwens J 2012 Kajian tentang sumber daya ikan dan kondisi lingkungan guna menentukan strategi pemanfaatan berkelanjutan di perairan Pulau Numfor *Laporan Akhir UPT Loka Konservasi Biota Laut-Biak* (Biak: UPT LKBL LIPI) p 32

[79] Suyarso dan Djuwariah 2008 *Studi baseline terumbu karang di lokasi Daerah Perlindungan Laut Biak, Kabupaten Biak Numfor* (Biak: COREMAP) p 120

[80] Lorwens J 2009 Kondisi ekosistem dan sumber daya pesisir Kabupaten Supiori, Provinsi Papua *Laporan Akhir UPT Loka Konservasi Biota Laut Biak – LIPI* (Biak: UPT LKBL LIPI) p 38

[81] Lorwens J 2013 Kajian tentang sumberdaya ikan karang dan kondisi lingkungan terumbu guna menentukan strategi pemanfaatannya secara berkelanjutan di perairan Aruri, Kabupaten Supiori, Propinsi Papua. *Laporan Akhir UPT Loka Konservasi Biota Laut – LIPI* (Biak: UPT LKBL LIPI) p 39

IOP Conf. Series: Earth and Environmental Science **184** (2018) 012010 doi:10.1088/1755-1315/184/1/012010

Assessment of tunicates found in the northern waters of Ambon Island and its rearing requirements in a laboratory setting

Y Tapilatu[1] and T Ginzel[1]

[1]Center for Deep-Sea Research, Indonesian Institute of Sciences (PPLD LIPI), Jl. Y. Syaranamual Guru-guru Poka Ambon 97233, Indonesia

yosmina.tapilatu@lipi.go.id

Abstract. The Banda Sea region is situated in the Wallacea biogeographical area, known as one of the biodiversity hotspots of the world. The region harbours various types of ecosystems, ranging from mangroves and coral reefs to seaweeds, indicating its potential for harbouring organisms that produce bioactive compounds, such as sea cucumbers, soft corals, and tunicates. While many studies have been carried out on the first two organisms, much lesser is known about the latter one, notably in this region. The assessment is our first attempt to explore the tunicates found in northern coastal waters of Ambon Island, an area that is a part of the Banda Sea. Two types of tunicates were obtained during the survey, namely the violet-green and white-yellow variants of solitary tunicates *Polycarpa aurata* (15 specimens) and social ones of *Didemnum molle* (3 specimens). The *Polycarpa* specimens' heights ranged from 2.5 to 4.8 cm with wet weights from 9.4 to 28.03 g, and the *Didemnum* ones were of 1.2 - 2 cm and wet weights from 1.2 to 5 g. Feeding experiment indicated that low volume (5-10 ml) of pure cultures such as *Dunaliella* sp., *Nannochloropsis* sp. and *Tetraselmis* sp. in small aquarium (40 x 60 cm) resulted in good survival rate for *Polycarpa aurata* collected from northern Ambon Island waters. Rearing experiments in a discontinuous aquarium setting using solitary tunicates indicated the following requirements: thorough cleaning of the tanks, regular feeding intervals, and regular cleaning of each tunicate. The regular cleaning of mucous at the external part of tunicates is indispensable to ensure the wellbeing of the organisms.

Keywords: Banda Sea, *Polycarpa* sp., discontinuous incubation, feeding experiment

1. Introduction

The Banda Sea is situated in the Wallacea biogeographical area, known as one of the biodiversity hotspots of the world. It harbors various types of ecosystems, ranging from mangroves and coral reefs to seaweeds, indicating the Banda Sea potential for harboring organisms that produce bioactive compounds, such as sea cucumbers, soft corals, and tunicates. While many studies have been carried out on the first two organisms, much lesser is known about the latter one, notably in this region. Northern coastal waters of Ambon Island are a part of the Banda Sea. Coral reefs coverage is relatively vast in these waters, but information regarding invertebrates living in them is scarce, if not known at all. This lack of information is not only caused by the limited studies carried out but also due to the lack of infrastructures to do such studies.

Tunicates are soft-bodied invertebrates (Class Ascidiacea) living in all sort of ecosystems, from coral reefs to the deep sea. They are very diverse with around 2,000 different species known to date.

Content from this work may be used under the terms of the Creative Commons Attribution 3.0 licence. Any further distribution of this work must maintain attribution to the author(s) and the title of the work, journal citation and DOI.
Published under licence by IOP Publishing Ltd

[1] They have various functions ranging from food sources in certain countries (Chile, South Korea, Japan and France) [2], producers of bioactive compounds [3-5] to biomarkers in ecotoxicological studies [6]. However, tunicates are much understudied in Indonesia, particularly in its eastern marine region. Literature search carried out online only yielded limited results on studies about and/or related to tunicates in Indonesian eastern marine region, such as tunicates inventory data from Siboga Expedition in 1899-1900 [7], a recent coral reef organisms inventory in Halmahera (2015) [8], the discovery of new alkaloids from *Polycarpa aurata* collected in north Sulawesi [4] and the identification of antimicrobial compound produced by a bacterial symbiont of *Phallusia julinea*, a tunicate species found in Doreri Gulf of West Papua [9]. None were available about incubation settings for tunicates found in the Banda Sea ecosystem. Moreover, there were articles available online about tunicates incubation in laboratory settings, but they were either too brief (and covered only temperate ones) [10] or only covered the social type of tunicate [11]. None to our knowledge existed on tropical tunicates incubation settings in limited infrastructures.

This assessment is thus our first attempt to explore the tunicates diversity in the Banda Sea. The main objective of this research is to provide the latest information about tunicates found in the said region and to provide initial information on its rearing possibility in a discontinuous aquarium setting for future research activities. The discontinuous setting is optimized for incubation since it is the only option available in a limited infrastructure.

2. Material and Method

2.1. Sample collection

Samples were collected from the northern coastal area of Ambon Island, in particular in three locations namely the waters of Morela, Mamala and Hitu villages (figure 1). These locations were selected due to their relatively preserved area of coral reefs. Tunicates were randomly collected using SCUBA diving equipment after visual assessment of its presence in the surveyed area. Sampling locations positions were recorded using a GPS Garmin 76 Cxi. Samples collected were put in sterile clipped plastic bags containing *in situ* sea water. Once reaching the surface the samples were kept in a cool box (8-10°C) throughout the sampling period until transfer to the aquaculture laboratory.

2.2. Sample identification

Tunicates obtained were identified based on its external appearance (body structure, tunic texture, etc). The identification was done according to guidelines that were available online, i.e. *Sous les Mers* [12], World Register of Marine Species (WoRMS) [13], Ocean Biogeographic Information System (OBIS) [14] and Reef Guide [15].

2.3. Incubation setting

Dimension and weight of samples obtained were measured and noted prior to transfer into the aquarium. Due to the limited infrastructure available at our aquaculture laboratory, we used only one aquarium in this experiment, with the dimensions of 40 x 60 cm. To mimic their original habitat, the tunicates were incubated with three sea cucumbers in the same aquarium. Natural seawater was pumped from nearby Ambon Bay and filtered prior to usage and replaced every two weeks. The bottom materials (sand and gravels) were replaced when algal growth was visible to eliminate the negative impact of organic waste build up and excessive algal growth in discontinuous incubation. Air was provided through a tube with constant pumping. Physico-chemical parameters were monitored at least once every two weeks to observe the water quality during incubation. Samples were observed daily and manual cleaning of mucous enveloping individual tunicates was done if necessary.

2.4. Feeding experiment

Tunicates obtained were fed with phytoplankton culture available at the laboratory, i.e *Nannochloropsis* sp., *Tetraselmis* sp. and *Dunaliella* sp. The phytoplankton cultures were fed in two treatments, namely individually (5 and 10 ml each) and the combination of three, 10 ml each (30 ml in total). All cultures were in their stationary phase by the time of feeding. Feeding was done every two weeks, and mortality rate would be recorded 7 days after feeding until cleaning and replacement of the aquarium, the sea water and the sand-gravel materials.

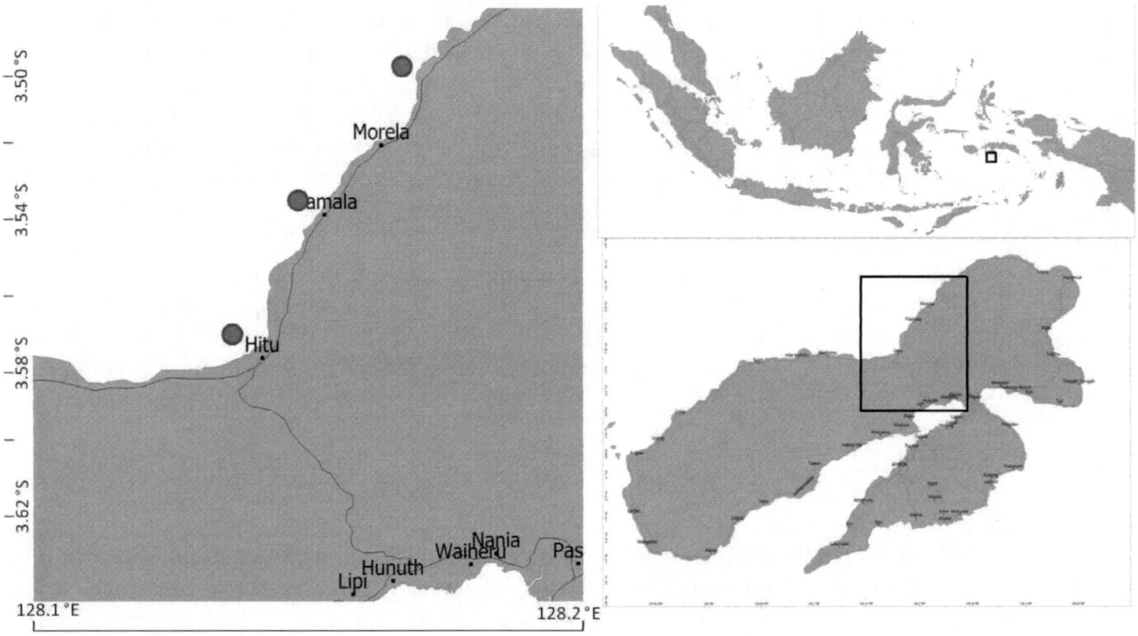

Figure 1. Map of Indonesia (inset, upper right), Ambon Island (inset, lower right) and tunicates sampling location (main, left). Blue dots indicate three locations where samples were collected.

3. Results and Discussion

3.1. Sampling locations profile

Blast fishing was a common practice in Lubang Buaya, Morela and adjacent area around 2005. It was done with the homemade bombs and caused important destruction in these waters. Fortunately, following the local authorities' intervention, the fishermen inhabiting the surrounding village did not continue their usage of this destructive method of fishing. Fragments of branched coral found were mostly due to anchoring and tourism activities. The coral reef condition of Lubang Buaya Morela, in general, was relatively good visually, with water transparency reached more than 75% of water depth (cf. Table 1) during the sampling and only a few floating plastic wastes observed. Overall Morela coral reef ecosystem was in better condition compared to those in Mamala and Hitumessing. The seabed in this coastal area was a shallow shelf from the intertidal area to about 50 meters seawards, continued by a steep slope.

Table 1. Sampling locations profiles.

Description	Morela	Mamala	Hitumessing
Position coordinates (Latitude, Longitude)	3°31′30.8″ S, 128°12′47.4″ E	3°33′16.5″ S, 128°11′26.1″ E	3°34′58.2″ S, 128°10′21.0″ E
Approximate depth (m)	6	4	6
Transparency (m)	6	3	5
Dominant substrate	Coral reef and sand	Sand	Muddy sand (close to river mouth)
Amount of samples collected	10	6	2

3.2. *Tunicate morphology and ecology*

There were 18 samples collected in total, with 15 of them were solitary tunicates and the other three were social type ones. Ten of fifteen *Polycarpa* sp. were collected from Morela, three from Mamala and two from Hitumessing area. Tunic morphology observation, height and distribution profile indicated solitary ones' resemblance to the violet-green and white-yellow variants of *Polycarpa aurata* (Quoy and Gaimard, 1834), commonly known as the ink-spot sea squirt [15], the ox-heart ascidian or the gold-mouth sea squirt. Their body forms were ovoid, globular, or conical, with the wrinkled leathery test. Their heights ranged from 2.5 to 4.8 cm with wet weights from 9.4 to 28.03 g. Their heights were slightly less compared to the available information in the identification guide (5 to 15 cm) [12]. They were numerous at the steep slope, notably in Morela area (Figure 2a). As suggested by Berrill in 1955, some species of tunicates from the family of Styelidae such as *Polycarpa* are characterized by the brittle sand-embedded test and muscles to withdraw apertures and flatten the body and are adapted to life on the open sea floor [16].

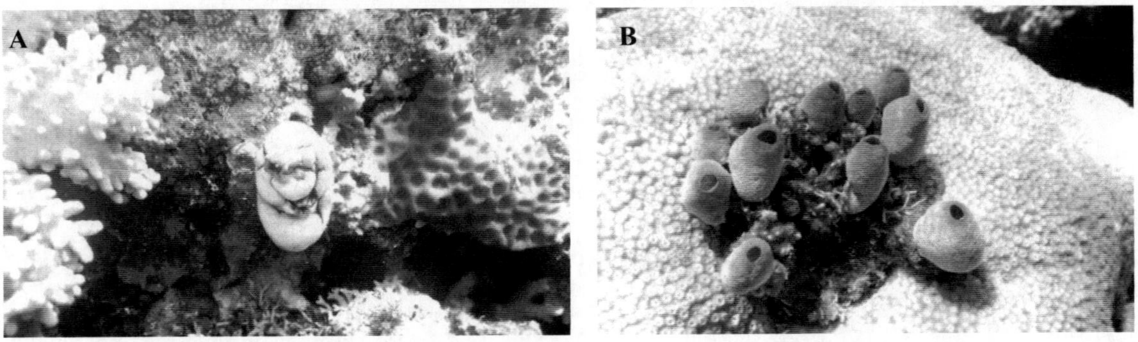

Figure 2. Solitary tunicate morphologically resembled *Polycarpa* sp. variant white-yellow found in the coral reef in Morela (a); a colony of *Didemnum* sp. as observed in Mamala (b). Both pictures were taken in early April 2017. Photo credit: Robert Alik.

The other three were *Didemnum molle* (Herdmann, 1886), a social type that was collected from Mamala area (Figure 2b), and commonly known as the green barrel sea squirt. The observed green and brown colors were from *Phrochlorons* sp., the cyanobacterial symbionts of *Didemnum* sp. [12]. The social type tunicates were visibly smaller and lighter, with height ranged from 1.2 to 2 cm and wet weights from 1.2 to 5.0 g only. The size of *Didemnum* sp. encountered in Mamala was smaller than reported elsewhere in Indonesia. In the Flores Sea, for example, the adult ones could reach 3 to 10 cm in height [12]. It is thus possible that the samples collected from Mamala were only young tunicates, or probably the surrounding environment did not facilitate the optimal growth of this tunicate, as was indicated by their limited occurrence in the three sampling areas.

Both tunicate types are commonly found in eastern Indonesian waters. They attach to hard substrates such as reefs and stones. *Polycarpa aurata* and *Didemnum molle* are found in the tropical eastern Indian Ocean and the western Pacific Ocean, including the Philippines, Indonesia and northern Australia. Their habitat depth ranges are 5 to 50 m (*Polycarpa*) and 5-30 m (*Didemnum*) [12]. A recent inventory of coral reef organisms recorded species of *Polycarpa* and *Didemnum* among ascidians found off western Halmahera [8], an island situated in the north Moluccas. However, further identification is necessary, either by dissection or genetically to confirm the affiliation of all samples collected in our study.

3.3. *Incubation in the discontinuous setting trial*

Incubation in the discontinuous setting of 40 x 60 cm aquarium (figure 3a) indicated solitary tunicates higher survival rate compared to the social type ones. In other parts of the world, certain species of *Didemnum* are included in invasive species list ([8] and references therein), due to its rapid growth and aggressive competing capability. However, the mortality rate of *Didemnum* collected was 100% by the third day of incubation in this setting. *Didemnum molle* was reported to live in the following range of natural environmental conditions: 19.13-28.95°C (temperature), 34.07-35.13 PSU (salinity) and at 3.56-4.78 ml l^{-1} (DO) [17].

Figure 3. Incubation setting in a 40 x 60 cm aquarium (a) and solitary tunicates in the aquarium, week 9 of incubation (b). *Polycarpa* sp. variant white-yellow (upper-center), violet-green (left-bottom) and sea cucumber (upper-right). The white arrow indicates the plastic pipe that was used to deliver air to the aquarium. The black arrows indicate opened siphons of a tunicate already adapted with the aquarium condition.

The small aquarium could be the cause of this high mortality rate, as the size did not permit incubation of too many tunicates, along with the sea cucumber. Other than that, the rapid growth of existing algae in the aquarium seawater resulted in the increase of water pH (table 2). This probably created a higher stress level to the social tunicates compared to the solitary ones and thus did not allow their optimal survival. The high fluctuation of salinity values could due to the evaporation or probably because the equipment used during the measurement was not properly calibrated.

Table 2. Water quality parameters observed in the aquarium. The values presented were obtained after six measurements during 13 weeks of incubation. Standard values were based on environmental conditions information of *Polycarpa aurata* according to the references cited.

Parameter	Measurements	Standard values for *Polycarpa aurata*	Standard reference
pH	8.24 ± 0.40	7.5-9	[10]
Salinity (PSU)	35.58 ± 4.25	34.98-35.13	[14]
Temperature (°C)	27.02 ± 0.50	26.22-26.8	[14]
DO (ml l^{-1})	6.6 ± 0.65	4.57-4.68	[14]

3.4. Feeding experiment

The feeding experiment carried out using *Polycarpa* sp. indicated that the highest mortality rate (40%) was observed when the tunicates were fed with 10 ml *Nannochloropsis* sp. (Table 3). The second highest mortality rate was observed when they were fed with 30 ml of combined three phytoplankton cultures, as it provoked rapid growth of algae at the aquarium wall and thus reduced the water quality. The combination also provoked rapid mucous envelopment of tunicates that led to mortality. Manual cleaning of enveloping mucous should be done at least once every two days to prevent mass mortality of tunicates.

Berrill [10] suggested the pure culture of *Nitzchia* sp. as the nutritional source, but mentioned that mixed assortment of small diatoms and algae could also be satisfactory. However, in our incubation settings, mixed culture tended to provoke a rapid increase of algal mass that occupied aquarium walls and thus increase pH values (cf. Table 2) that could provoke mortality. The lower dose (5-10 ml) of pure cultures such as *Dunaliella* sp., *Nannochloropsis* sp. and *Tetraselmis* sp. indicated good survival rate for *Polycarpa aurata* collected from northern Ambon Island waters and incubated in small aquarium. Nevertheless, this should be confirmed in the future with further experiments where the aquarium is kept in more sheltered place for incubation setting optimization.

Table 3. Feeding experiment and mortality rate of tunicates (*Polycarpa* sp.) from northern Ambon Island waters. The incubation was carried out for 7 days before the sea water was replaced with the clean one.

No.	Species	Feeding experiment Volume per species (ml)	Total volume fed	Mortality rate (%)
1	*Dunaliella* sp.	5	5	0
2	*Dunaliella* sp.	10	10	29
3	*Nannochloropsis* sp.	5	5	0
4	*Nannochloropsis* sp.	10	10	40
5	*Tetraselmis* sp.	5	5	22
6	*Tetraselmis* sp.	10	10	0
7	*Dunaliella* sp.	10		
	Tetraselmis sp.	10	30	33
	Nannochloropsis sp.	10		

4. Conclusions

The assessment is our first attempt to explore the tunicates found in northern coastal waters of Ambon Island, an area that is a part of the Banda Sea. Two types of tunicates were obtained during the survey, namely the violet-green and white-yellow variants of solitary tunicates *Polycarpa aurata* and social ones of *Didemnum molle*. Rearing experiments in a discontinuous aquarium setting using solitary tunicates indicated the following requirements: thorough cleaning of the tanks, regular feeding intervals, and regular cleaning of each tunicate. The regular cleaning of mucus at the external part of tunicates is indispensable to ensure the well-being of the organisms.

Future experiments should include incubation in a more sheltered setting, with only a certain amount of tunicates (8 to 10 solitary ones) in the same dimension of the aquarium. Further identification to confirm samples' identity would be necessary. More importantly, bioprospecting effort could be initiated to examine whether *Polycarpa aurata* growing in Ambon island waters contain similar bioactive compounds as those sampled in Manado in a prior report.

Acknowledgement

This research was supported financially by LIPI's internal funding for research, as part of the research project on sustainable utilization of marine bioresources in Ambon Island northern coastal waters. We thank Abdul Wahab Radjab, the coordinator of the project, to facilitate the sampling trip; Robert Alik, Terry Indrabudi and Arahman for their help during sampling and Wempy Barends for the provision of phytoplanktons used in this study. We gratefully acknowledge the critical reviews by two anonymous reviewers that help improve the quality of this manuscript. This is contribution number 0011 from Marine Microbiology and Biotechnology Laboratory, Centre for Deep-Sea Research LIPI.

References

[1] Ruppert E E, Fox R S and Barnes R D 2004 *Invertebrate Zoology* 7th ed (Boston: Cengage Learning) pp 940–956

[2] Lambert G, Karney R C, Rhee W Y and Carman M R 2016 Wild and cultured edible tunicates: a review *Manag. Biol. Invasions* **7**(1) 59–66

[3] Suarez-Jimenez G M, Burgos-Hernandez A and Ezquerra-Brauer J M 2012 Bioactive peptides and depsipeptides with anticancer potential: Sources from marine animals *Mar. Drugs* **10**(5) 963–86 doi: 10.3390/md10050963

[4] Wang W, Oda T, Fujita A, Mangindaan R E P, Nakazawa T, Ukai K, Kobayashi H and Namikoshi M 2007 Three new sulfur-containing alkaloids, polycarpaurines A, B, and C, from an Indonesian ascidian *Polycarpa aurata Tetrahedron* **63**(2) 409–12 doi: 10.1016/j.tet.2006.10.060

[5] Oku N, Matsunaga S and Fusetani N 2003 Shishijimicins A–C, Novel Enediyne Antitumor Antibiotics from the Ascidian *Didemnum proliferum J. Am. Chem. Soci.* **125**(8) 2044–5 http://dx.doi.org/10.1021/ja0296780

[6] Parrinello D, Bellante A, Parisi M G, Sanfratello M A, Indelicato S, Piazzese D and Cammarata M 2017 The ascidian *Styela plicata* hemocytes as a potential biomarker of marine pollution: In vitro effects of seawater and organic mercury *Ecotoxicol. Environ. Saf.* **136** 126–34 doi: 10.1016/j.ecoenv.2016.11.001

[7] Sluiter C P 1904 Die Tunicaten der Siboga-Expedition: Die Socialen und Holosomen Ascidien (*Siboga Expeditie* 1899-1900 vol L17) ed M Weber (Leiden: Brill) 208 p

[8] Gittenberger A, Draisma S G A, Arbi U Y, Langenberg V, Erftemeijer P L A, Tuti Y and Hoeksema BW 2015 Coral reef organisms as bioregion indicators off Halmahera, Moluccas, Indonesia *Aquat. Conserv. Mar. Freshw. Ecosyst.* **25**(6) 743–55.

[9] Mogea R A, Suharjono, Ardyati T and Samino S 2015 Identification of antimicrobial compounds produced by *Pseudomonas aeruginosa* NS3 symbion ascidian *Phallusia julinea Int. J. ChemTech Res.* **8**(3) 1036–40 ISSN: 0974-4290

[10] Berrill N J 1937 Culture methods for ascidians (*Culture methods for invertebrate animals*) eds P S Galtsoff, F E Lutz, P S Welch and J G Needham (New York: Comstock Publishing Co.) pp 564-571

[11] Epelbaum A, Therriault T W, Paulson A and Pearce C M 2009 Botryllid tunicates: Culture techniques and experimental procedures *Aquat. Invasions* **4**(1) 111–20

[12] Cornu F 2009 *Sous Les Mers* [Internet] accessed Aug 1 2017 from www.souslesmers.fr

[13] WoRMS Editorial Board 2017 *World Register of Marine Species* [Internet] accessed Aug 1 2017 from http://www.marinespecies.org at VLIZ. doi:10.14284/170

[14] AIMS Bioresources Library 2017 *Polycarpa aurata* (Quoy & Gaimard, 1834) *Ocean Biogeographic Information System* [Internet] accessed Aug 1 2017 from http://iobis.org/explore/#/taxon/495246

[15] Charpin F 2017 Ink-Spot Sea Squirt (*Guide to the Tropical Reefs*) [Internet] accessed Aug 1 2017 from https://reefguide.org/

[16] Kott P 2005 *Catalogue of Tunicata in Australian waters* (Brisbane: Queensland Museum) 305 p

[17] AIMS Bioresources Library 2017 *Didemnum molle* (Herdmann, 1886) *Ocean Biogeographic Information System* [Internet] accessed Aug 1 2017 from http://iobis.org/explore/#/taxon/434288

Physical properties and nutrients distribution of seawater in the Banda Sea - Indonesia

M L Firdaus[1]

[1]Graduate School of Science Education, University of Bengkulu, Bengkulu 38371, Indonesia

E-mail: lutfi@unib.ac.id

Abstract. The Banda Sea is a sea in the eastern part of the Indonesian archipelago, connected directly to the Pacific Ocean and surrounded by hundreds of small islands that create its unique characteristics. A large portion of Pacific seawater is passed through the Banda Sea to the Indian Ocean forming an Indonesian through flow (ITF), an ocean current with importance for global climate. The ITF provides a low-latitude pathway for warm, fresh water to move from the Pacific to the Indian Ocean and serves as the upper branch of the global heat conveyor belt. Despite its importance in the global ocean and climate system, detail study and discussion of the Banda Sea are scarce. Here, we discuss the physical properties in term of temperature, salinity and density in conjunction with nutrients distribution in seawater from 17 stations of the Banda Sea using the direct observation data of the research cruises that were gathered by Reid and Mantyla. The available numerical data of the Banda Sea were then visualized using Ocean Data View software. To the best of our knowledge, it is the first report for a detailed discussion of this kind in the Banda Sea. The results show that nutrients concentration depleted in surface water (0 – 500 m) and enriched in deep water (500 m – bottom). The concentrations ranges of phosphate, nitrate and silicate were 0.3 – 3.1, 0.5 – 40 and 1 – 147 μmol l^{-1}, respectively. As shown from temperature, salinity and density profiles, it is known that the inflowing seawater from the Pacific Ocean into the Banda Sea is gradually modified which change the upper water masses (0 – 250 m) characteristics that eventually outflows into the Indian Ocean.

Keywords: Banda Sea, nutrient, salinity, temperature, oxygen.

1. Introduction

Indonesian through flow is governed by a strong pressure gradient between seawater in the western Pacific and the eastern Indian Ocean. The average sea level difference is 16 cm and the most of the pressure gradient is contained in the upper 200 m. The annual pressure gradient signal has a maximum in July and August during the southeast monsoon and a minimum in January and February during the northwest monsoon [1]. The interoceanic transfer of tropical seawater from the Pacific to the Indian Ocean through the Indonesian seas carries a substantial amount of seawater and heat that linked to the climate system such as the Asian Monsoon, the Indian Ocean Dipole and El Nino Southern Oscillation. Based on recent observation data, the flux of this seawater mass transfer was as high as 15 Sv (15×10^6 m^3 s^{-1}) which is about ten times greater than all of the river flow of the Earth [2].

Content from this work may be used under the terms of the Creative Commons Attribution 3.0 licence. Any further distribution of this work must maintain attribution to the author(s) and the title of the work, journal citation and DOI.
Published under licence by IOP Publishing Ltd

The inflows Pacific Ocean seawater stays for ~1 year in the Banda Sea [3] before it outflows to the Indian Ocean. During its stay, the seawater is modified within the Banda Sea by mixing, with energy derived from powerful tidal currents, by Ekman pumping, as well as heat and freshwater flux across the sea–air interface. This results in a unique Indonesian tropical stratification with a strong, although relatively isohaline, thermocline. The modified seawater has a major impact on the heat content and sea surface temperature patterns of the Indian Ocean [2].

In contrast to the importance of the Banda Sea on world ocean and global climate system, detail report on physical properties and nutrients concentration of Banda Sea seawater are limited. Here we discuss the previously collected research cruises data in term of temperature, salinity, oxygen, phosphate and silicate concentration. The discussion in this paper will enrich the knowledge of particularly Indonesian oceanographers on the Banda Sea characteristics.

2. Materials and Methods

Sampling locations and data acquisition

Indonesian archipelago and the Banda Sea are located between western equatorial Pacific Ocean and the eastern north Indian Ocean. Seawater samples were collected from several research cruises that provided open-access cruises data. The data were collected in October, 2000. These results were then gathered and organized by Reid and Mantyla [4]. The metadata is available freely from NOAA (Figure 1) and Ocean Data View data center. In this paper, we selected and processed 17 stations in the Banda Sea. The deepest bottom water was in station 8 (7,025 m depth) which is then followed by station 1. Hydrographic and nutrients data are shown in Table 1. Due to an extensive numerical data from these 17 stations yet limited space usable, we can only show a small portion of the data here. The rest of the data are available upon request to the author.

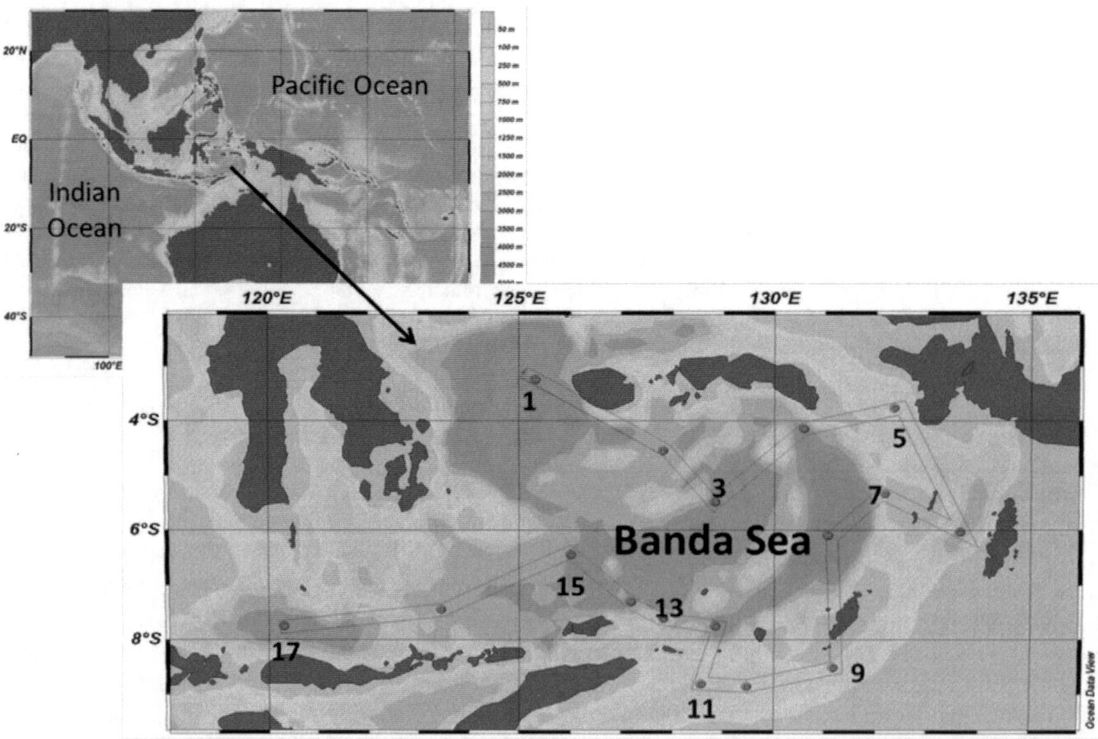

Figure 1. Location of sampling stations in the Banda Sea. The designated station name and section are shown on the map. The plot was made using Ocean Data View [5].

Table 1. Physical properties and nutrients concentration

Cruise	Station	Longitude [°East]	Latitude [°North]	Depth [m]	Temp [°C]	Salinity [psu]	Pot.Density [kg m^{-3}]	Oxygen [ml l^{-1}]	Phosphate [µmol l^{-1}]	Nitrate [µmol l^{-1}]	Silicate [µmol l^{-1}]
31WT	4	130.58	-4.15	1	26.47	34.47	22.49	5	0.24	1.1	3
				9	26.35	34.47	22.52	5.12	0.24	1.1	3
				19	26.13	34.46	22.59	5.06	0.24	1.1	3
				30	25.24	34.34	22.78	4.84	0.28	1.9	3
				49	23.35	34.41	23.39	3.61	0.64	8.5	9
				74	21.9	34.52	23.89	3.3	0.89	11.5	14
				98	20.1	34.57	24.41	3.11	1.01	13.7	18
				123	19.57	34.56	24.55	3.05	1.03	14.6	19
				148	18.07	34.59	24.95	2.98	1.18	16.2	22
				198	16.37	34.64	25.40	2.7	1.43	19.6	29
				247	12.98	34.61	26.11	2.61	1.73	24.3	39
				297	11.18	34.59	26.43	2.45	2	27.2	46
				397	9.07	34.58	26.79	2.35	2.22	30.8	58
				496	7.79	34.59	27.00	2.36	2.35	32.8	69
				597	6.89	34.59	27.13	2.3	2.5	34.1	78
				696	6.02	34.59	27.24	2.28	2.55	35.2	88
				796	5.43	34.60	27.32	2.32	2.59	35.9	95
				996	4.79	34.60	27.40	2.35	2.67	36.8	105
				1196	4.09	34.61	27.48	2.39	2.72	37.7	116
				1494	3.57	34.61	27.54	2.41	2.78	38.2	128
				2954	3.066	34.62	27.61	2.49	2.89	38.7	137
				3249	3.082	34.62	27.61	2.49	2.88	38.7	138
				3545	3.108	34.62	27.61	2.53	2.89	38.7	136
				3839	3.14	34.62	27.61	2.47	2.89	38.9	137
31WT	7	132.17	-5.3333	0	26.52	34.53	22.52	5.05	0.24	0.3	3
				10	26.08	34.53	22.66	4.99	0.25	0.3	3
				20	26.1	34.61	22.71	5.12	0.23	0.2	2
				30	26.08	34.64	22.74	5.03	0.21	0.3	3
				49	22.84	34.51	23.61	3.34	0.84	10.5	11
				65	21.64	34.53	23.97	3.1	1	12.6	15
				74	20.92	34.55	24.18	3	1.06	13.6	16
				98	19.3	34.63	24.67	2.78	1.24	16.4	21
				123	18.25	34.65	24.95	2.68	1.36	17.9	23
				149	17.77	34.73	25.13	2.67	1.43	18.7	25
				198	15.72	34.70	25.59	2.55	1.61	21.4	31
				247	13.35	34.64	26.06	2.35	1.88	25	41
				297	11.44	34.61	26.41	2.25	2.08	27.8	50
				347	9.96	34.59	26.65	2.26	2.18	29.9	56
				397	9.08	34.59	26.80	2.28	2.35	31.2	61
				447	8.3	34.59	26.92	2.19	2.42	32.4	67
				547	7.12	34.59	27.09	2.19	2.56	34.1	78
				649	6.78	34.59	27.14	2.19	2.62	34.6	83
				751	6.4	34.59	27.20	2.15	2.68	35	87

3. Results and Discussion

Figure 2 and 3 show some physical properties of the Banda Sea seawater. The surface temperature of seawater was about 26°C with thermocline occurred from 50 to 500 meter depth. Seawater temperature was relatively uniform from 1,400 meter to the bottom. The lowest temperature of seawater in bottom water was about 3.1°C. Most of seawater has a density more than 27 kg m^{-3} that occupy the water column from 500 meter to the bottom. The concentration of oxygen was high (~ 4.86 ml l^{-1}) in surface water and decreasing toward the deep water (~ 2.35 ml l^{-1}), directly proportional to the density and water masses, that shows effective ventilation in the Banda Sea [3].

Water mass on the deepest depth with temperature 3 – 10°C and salinity 33.8 – 34.8 was characterized as Antarctic Intermediate Water (AAIW). This water mass that originally occupied the Pacific Ocean seawater from 500 to 1,500 meter depth is surprisingly occupied bottom water of the Banda Sea. Station 8 has the deepest bottom water that is even deeper than the Pacific Ocean in average (see Figure 1 and 5). However, all of the bottom waters in the Banda Sea were not occupied by Antarctic Bottom Water (AABW) that has a very low temperature (about 2°C). It is likely that the Banda Sea is partly blocked by shallow ocean floor of the surrounding islands that made it as a marginal sea such as Lifamatola passage that has a ~2,000 m sill depth. Therefore, AABW that occupied most of deep water of the Pacific Ocean was not able to enter the Banda Sea [6-7].

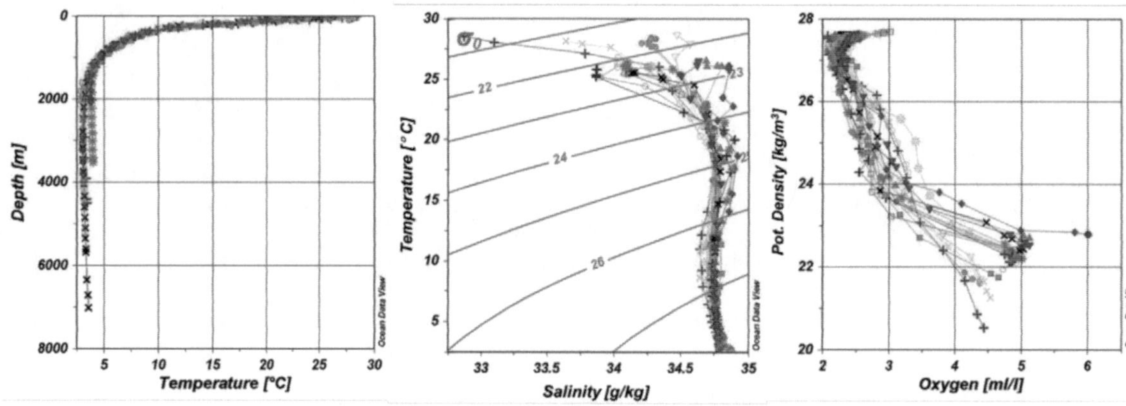

Figure 2. Vertical distribution of temperature showing a thermocline profile (left). Temperature-Salinity (T-S) curve with isopycnal contours showing distinct water masses that occupied the Banda Sea (center). Oxygen concentration as a function of density is plotted on the right panel.

As shown in Figure 3, the AABW was overlaid by North Pacific Intermediate Water (NPIW) water mass (3.5–15°C, 34.0– 34.6%). Indonesian Intermediate Water (IIW) (3.5–20.5°C, 34.6– 34.7%) occupied the intermediate seawater of the Banda Sea, while surface water was dominated by Indonesian Upper Water (IUW) (8.0–23.0°C, 34.4–35.0%). In contrast, the adjacent western equatorial Pacific Ocean seawater that flow to Indonesian Seas has a more saline and colder seawater characterized as Pacific Equatorial Water (PEW; 7.0–23.0°C, 34.5–36.0%) and Western North Pacific Central Water (WNPCW; 10.0–22.0°C, 34.2–35.2%). Freshwater inputs from Indonesian archipelago are responsible for this salinity dilution. These water masses characterization is consistent with previous reports [5-6, 8–10]. Eventually, the modified inflow Pacific Ocean seawater (broken blue circle) is outflow to the Indian Ocean that has similar TS characteristics with those of Banda Sea seawater (broken red circle, Figure 3).

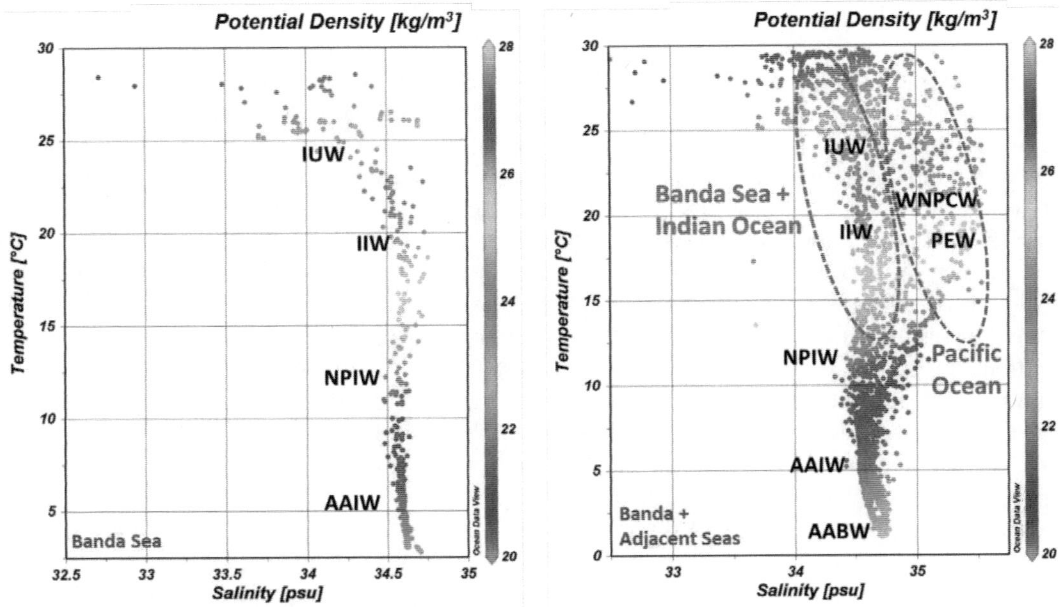

Figure 3. The T-S diagram shows distinct water masses that occupied the Banda Sea with density in color gradation (Left). Data from adjacent Pacific and Indian Ocean seawater are added to those of Banda Sea (right). The identified water masses are Antarctic Intermediate Water (AAIW), North Pacific Intermediate Water (NPIW), Indonesian Intermediate Water (IIW) and Indonesian Upper Water (IUW) with additional Antarctic Bottom Water (AABW), Pacific Equatorial Water (PEW) and Western North Pacific Central Water (WNPCW) in the Pacific Ocean. See the text for more detail discussion.

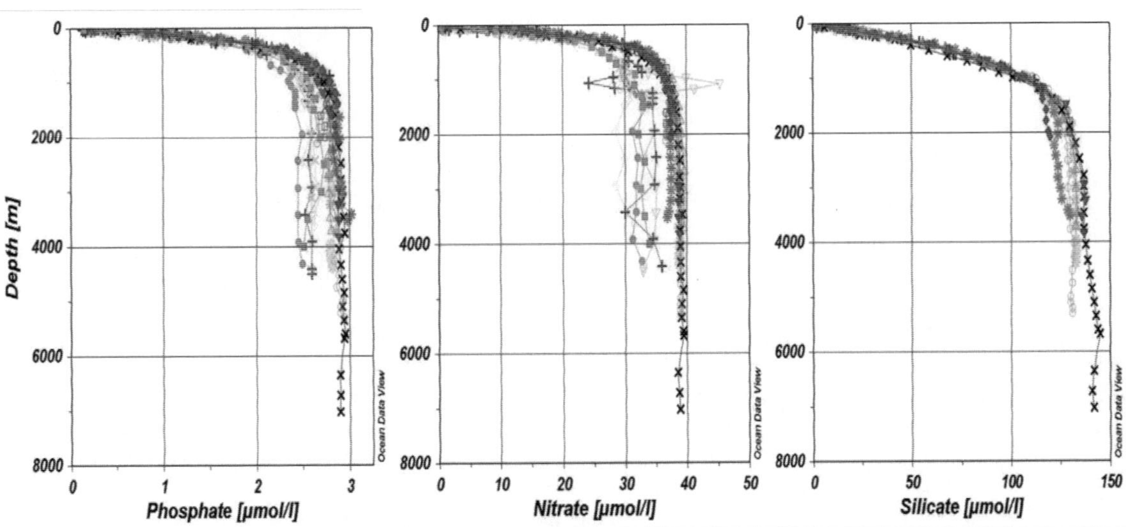

Figure 4. Vertical profiles of nutrients in the Banda Sea.

Figure 4 and 5 show nutrients concentration in the Banda Sea. Phosphate, nitrate, and silicate were depleted in surface water due to the uptake by phytoplankton and were enriched in deep water due to the recycling and remineralization from sinking particulates. These three compounds were classified as macro nutrients which are essential for phytoplankton growth [11-12]. Concentrations of these nutrients in the upper water of the Banda Sea were higher (5 – 60%) than those in the adjacent Pacific Ocean. These high concentrations of nutrients probably occurred due to the effective upwelling of deep seawater and nutrients input from terrestrial sources within the Banda Sea [11-14] especially during the (southeast) monsoon. Previous results also show the high concentration of trace metals as micronutrients in Indonesian seawater due to terrestrial and riverine fluxes, seasonal wind-induced upwelling, and/or large tidal forces [14]. Slightly higher concentrations of nutrients (i.e. phosphate) occurred at stations 1, 5, 6, 7, 8 and 15 (Figure 5). Westeyn *et al.* (1990) analyzed nutrients in the Banda Sea during and after upwelling season. Phosphate, nitrate and silicate concentrations were from 0.04 to 2.05, 0 to 28.1 and 1.0 to 47.3 μmol l^{-1}, respectively. Therefore, the present study shows relatively coherent nutrients concentration with those of Westeyn *et al.* [15] . The high nutrients concentrations were maintained throughout the outgoing flow into the Indian Ocean as can be seen on the nutrients profiles in the adjacent Indian Ocean seawater (data are not shown here). In the eastern Indian Ocean close to the outgoing ITF site, e.g. station 7B (115.38°E; 13.48°S), concentrations of phosphate and nitrate at depth 668 m were 2.27 and 35.1 μmol l^{-1}, respectively. In contrast, in the western Pacific Ocean close to inflow of ITF, e.g. station RY5860 (132°E; 2°N) nutrients of phosphate, nitrate, and silicate concentrations at depth 451 m were 2.06, 28.0, and 131 μmol l^{-1}, respectively.

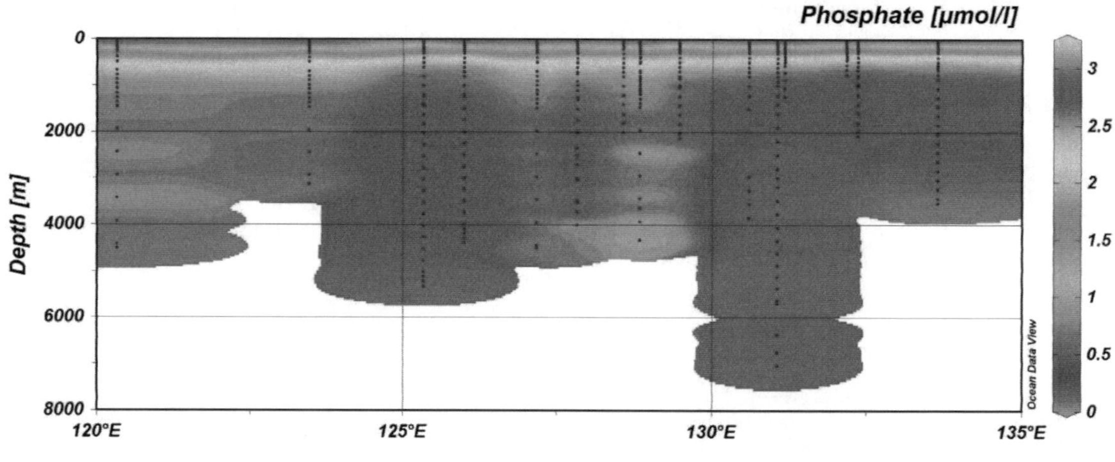

Figure 5. Sectional distribution of phosphate (μmol l^{-1}) in the Banda Sea. The black dots indicate the point of sampling.

4. Conclusion

Based on the temperature, salinity and density profiles, we identified at least four major water masses that occupied the Banda Sea, i.e. Antarctic Intermediate Water (AAIW), North Pacific Intermediate Water (NPIW), Indonesian Intermediate Water (IIW) and Indonesian Upper Water (IUW). It is also known that seawater inflows from the Pacific Ocean into the Banda Sea are gradually modified which change the upper (0 – 250 m) water masses characteristics (e.g. less saline and warmer) that continually outflows into the Indian Ocean. The results indicated that nutrients concentration in the Banda Sea depleted in surface water and enriched in deep water. The concentrations range from surface to bottom water of phosphate, nitrate and silicate were 0.3 – 3.1, 0.5 – 40 and 1 – 147 μmol l^{-1}, respectively. The surface water of the Banda Sea has higher concentrations of nutrients than those in the adjacent Pacific Ocean seawater, and maintained through their outgoing into the Indian Ocean. These results confirm that the Banda Sea takes an essential part of the global thermohaline circulation which eventually controls the global climate system.

Acknowledgements
We are grateful for the financial support from the Indonesia Ministry of Research, Technology and Higher Education (Kemenristek-Dikti). We would like to thank anonymous reviewers for their valuable comments on the manuscript.

References

[1] Wyrtki K 1987 *J. Geophys. Res.* **92** 941
[2] Gordon A, Sprintall J, Van Aken H M, Susanto D and Wijffels S 2010 *Dyn. Atmos. Ocean.* **50** 115
[3] Hautala S L, Reid J L and Bray N 1996 *J. Geophys. Res.* **101** 12375
[4] Reid, J. 2013 *Java Ocean Atlas - Reid/Mantyla Section Data, a Library of More than 2000 Oceanographic Sections Developed from the Cruises Used in the Reid/Mantyla Pre-WOCE Data Set* (San Diego: University of California - Scripps Institution of Oceanography)
[5] Schlitzer R 2015 *Ocean Data View, https://odv.awi.de/*
[6] Emery W J 2001 *Water Types and Water Masses in Encyclopedia of ocean sciences* (London: Academic Press)
[7] Firdaus M L, Minami T, Norisuye K and Sohrin Y 2011 *Nat. Geosci.* **4** 227
[8] Ilahude A G and Gordon A 1996 *J. Geophys. Res.* **101** 12401
[9] Gordon A L and Susanto R 2001 *Ocean Dyn.* **52** 2
[10] Firdaus M L, Norisuye K, Nakagawa Y, Nakatsuka S and Sohrin Y 2008 *J. Oceanogr.*, **64** 247
[11] Gieskes W W C, Kraay G W, Nontji A, Setiapermana D and Sutomo 1988 *Netherlands J. Sea Res.* **22** 123
[12] Nakatsuka S, Okamura K, Takeda S, Nishioka J, Firdaus M L, Norisuye K and Sohrin Y 2009 *Deep Sea Res. Part II* **56** 2822
[13] Arinardi O H, Baars M A and Oosterhuis S S 1990 *Netherlands J. Sea Res.* **25**, 545
[14] Firdaus M L, Darti P, Alwi W, Swistoro E, Sundaryono A and Ruyani A 2015 *AIP Conf. Proc.*, **1677** 110003
[15] Westeyn F J, Ilahude A G, Baars M A 1990 *Netherlands J. Sea Res.* **25** 449

IOP Conf. Series: Earth and Environmental Science **184** (2018) 012012 doi:10.1088/1755-1315/184/1/012012

Possible impact of El Niño and La Niña on water mass circulation in Ambon Bay

F R T Saputra[1], M R Putri[2] and W J Tattipata[1]

[1]Centre for Deep-Sea Research, Indonesian Institute of Science (PPLD LIPI), Jl. Syaranamual Guru-Guru Poka Ambon 97233, Indonesia
[2]Research Group for Oceanography, Institut Teknologi Bandung, Jalan Ganesa 10, Bandung 40132, Indonesia

fril001@lipi.go.id

Abstract. El Niño and La Niña are disruptions to the ocean-atmosphere system in the equatorial Pacific area, with significant impacts on the climate and weather throughout the world. This paper discusses the possible impact of El Niño and La Niña to the water mass circulation in Ambon bay which is directly adjacent to the Banda Sea. We used observed vertical temperature variability data in June, July, September, and October in the year of 2008 (neutral year) and 2009 (El Niño year), and July and October 2010 (La Niña year). All data have been taken with CTD instrument in 7 stations which are aligned in one straight line from inner to Ambon Outer Bay (AOB). The result shows that the water mass circulation in Ambon Bay may be affected by El Niño and La Niña. El Niño phase makes the thermocline layer from the Banda Sea to become shallower than usual and thus makes the circulation in Ambon Bay more rigorous. This is because the water mass with higher density can enter into the deep Ambon Inner Bay (AIB) through the narrow sill which separates the Ambon Bay inner from outer parts. La Niña, on the other hand, weakens the water circulation because the thermocline layer becomes deeper than usual.

Keywords: El Niño, La Niña, water mass circulation, Ambon Bay

1. Introduction

The dynamics of Indonesian Waters are affected by Indonesia's geographical position which lies between two oceans, namely the Pacific and Indian Oceans. In Eastern Indonesia, the Indonesian throughflow (ITF) advects water masses from the Pacific to Indian Oceans [1]. Some researchers show that sea surface temperature and heat in the Pacific and Indian Oceans are influenced by the ITF. The sea surface will rise in the Pacific Ocean and decrease in the Indian Ocean as much as 2 to 10 cm if ITF did not exist [2]. Water mass structure in Indonesian Waters is commonly affected by the characteristic of water mass from the Pacific and monsoon wind system [3].

The appearance of El Niño and La Niña in the tropical Pacific impacts the ocean dynamics and weather in Indonesia [4]. El Niño and La Niña arise through ocean-atmosphere interaction along the equatorial Pacific Ocean, resulting in deviations of ocean and atmospheric circulation from the normal condition. National Oceanic and Atmospheric Administration (NOAA) defines El Niño (warming phase) as an increase in sea surface temperature in Niño 3.4 region (east-central equatorial Pacific between 120°-170°W) above 0.5°C over successive 3 months. Meanwhile, La Niña (cooling phase) is

Content from this work may be used under the terms of the Creative Commons Attribution 3.0 licence. Any further distribution of this work must maintain attribution to the author(s) and the title of the work, journal citation and DOI.
Published under licence by IOP Publishing Ltd

defined as a reduction in sea surface temperature in Niño 3.4 region by more than 0.5°C over successive 3 months. The impact caused by El Niño and La Niña have different characteristics in different locations, in association with their pattern, intensity and duration [5-7].

In Indonesia, the appearance of El Niño causes the sea surface temperature to become lower than normal because the warm water pool which is located in the Indonesian waters migrates eastward toward the Central Pacific. Meanwhile, the appearance of La Niña causes sea surface temperature to be higher than usual as warm water is piled up toward the Maritime Continent [8,9]. Observation for temperature variability and chlorophyll-a had been done in some locations in Indonesia such as Bali and Sunda Strait [10,11] and southern of Java Island through the Timor Sea [3].

Ambon Bay which is located in the Banda Sea ecosystem has a unique characteristic; this bay is separated into two parts, Ambon Inner and Outer Bay. Ambon Inner Bay (AIB) is a semi-enclosed part of the bay with depth of only ±45 m, whereas the depth in Ambon Outer Bay (AOB) can reach ± 600m at the tip of the bay and directly facing the Banda Sea. There is a narrow sill that separate Ambon Inner and Outer Bay with depth of only ±15m; this sill limits water mass circulation in Ambon Bay [12]. Water mass from thermocline layer in the Banda Sea shoals and can reach ±70 m in depth when upwelling occurs in southeast monsoon [13], and is predicted to enter the AIB through tidal upwelling mechanism [14]. Water mass density from the Banda Sea is higher than water mass in AIB so it is predicted that a flushing phenomenon would occur in Inner Ambon Bay during southeast monsoon [15,16]. Water mass circulation has an important role in enriched water environment in AIB because it makes nutrient in the bottom to be transported toward the surface. Ambon Bay is known as one of the potential areas to bait fish farm. The changes in the marine environment in Ambon Bay will affect all marine organisms that live in it and eventually may affect fishing activities in the Banda Sea ecosystem.

We predicted that water mass circulation in Ambon Bay will be affected by El Niño and La Niña due to changes in the thermocline depth in the Banda Sea. Therefore, this study observes the effect of El Niño and La Niña in the southeast monsoon to water circulation through vertical temperature variability in Ambon Bay.

2. Methodology

In this observation, we use vertical temperature variability and ocean current data taken from monitoring activity of Ambon Bay which was conducted by the Ambon Technical Implementation Unit of Marine Biota Conservation Office LIPI (UPT BKBL LIPI, later known as the Centre for Deep -Sea Research, LIPI) from 2008 until 2010 in order to reveal the effect of El Niño and La Niña on water mass circulation in Ambon Bay. The main focus was on the southeast monsoon when flushing phenomenon usually occurs in the deep AIB.

Because of lack of data, we only use vertical temperature variability data in June, July, September and October 2008 and 2009, and also data in July and October 2010. The data was taken in seven stations along a straight line from AIB to AOB (figure 1) using CTD SBE Seabird 19 v2 measurements and processed by ODV (Ocean Data View) software [17]. We also use ocean current data measured on 18-19 July 2009 (24 hours observation) using current meter ALEC Compact – EM Model AEM-HR located in station 3 (1m, 5m, 30m, and 35m in depth). Observation through modelling to reveal the linkage of ENSO and monsoon with the sea surface temperature variability in Ambon Bay from 2000 until 2008 indicated that the sea surface anomaly in Ambon Bay was dominantly affected by local atmospheric condition rather than ENSO [18].

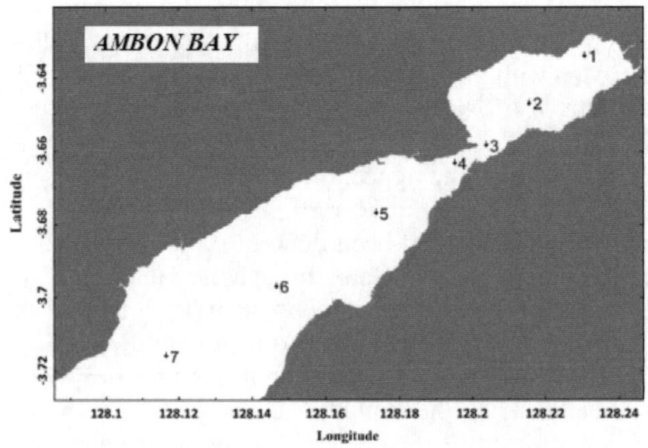

Figure 1. Map of observation location

3. Result and Discussion

Figure 2 shows the tidal prediction in Ambon Bay on 18-19 July 2009 [19], while figure 3 shows a 24-hour current measurement at station 3 starting from 16.00 pm 18 July 2009. The result shows that ocean current at the surface tends to have a different pattern with the current at deep, due to surface winds.

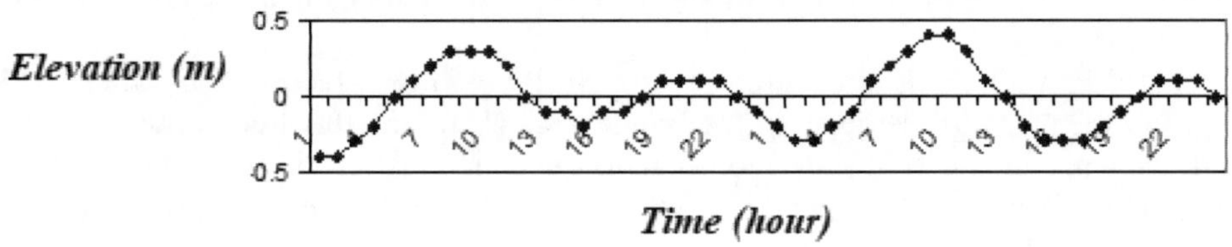

Figure 2. Tidal prediction for Ambon Bay, 18 – 19 July 2009

On 18 July 2009 at 18:00 to 24:00 local time (high tide), the ocean current at 5, 30 and 35 m depth flows into AIB (negative velocity). In contrast, the surface current flows to AOB (positive velocity). Furthermore, on July 19, 2009 at 00:00 to 06:00 am (low tide), the ocean current outflow has a maximum velocity of 0.4 $m.s^{-1}$. We predict the ocean current at 30 and 35 m depth will progress upward to the surface when it is blocked by the narrow sill at station 4. Some of this current will go out to AOB, but most of it is to turn left and right back into the AIB.

The distribution of ocean current in Ambon Bay is more dominated by tidal period rather than seasons. The influence of seasons on ocean currents is only visible in the surface layer, meanwhile the influence of tidal are more dominant in the deeper layer.

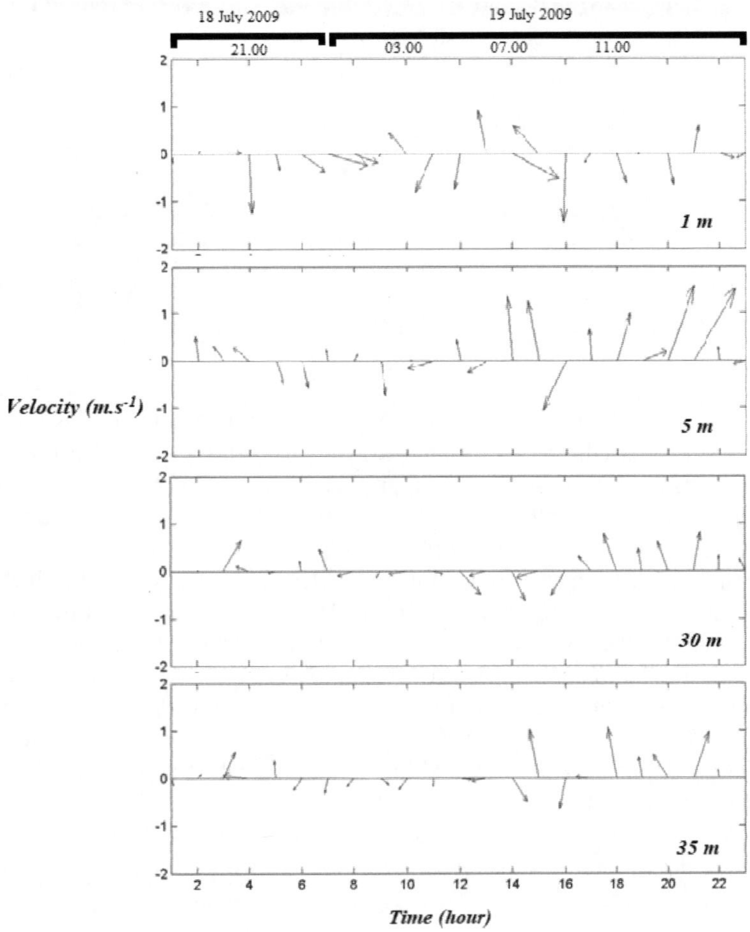

Figure 3. Ocean current in station 3

In this observation, the discussion about El Niño and La Niña refers to Niño 3.4 indices for the year 2008, 2009, and 2010 [20]. Based on figure 4, we can categorize from June to October 2008 as normal condition whereas from June to October 2009 as El Niño event, and June to October 2010 as La Niña event.

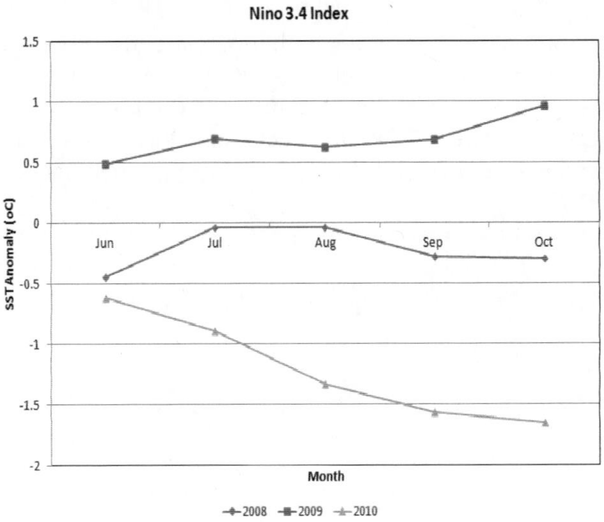

Figure 4. Niño 3.4 Index

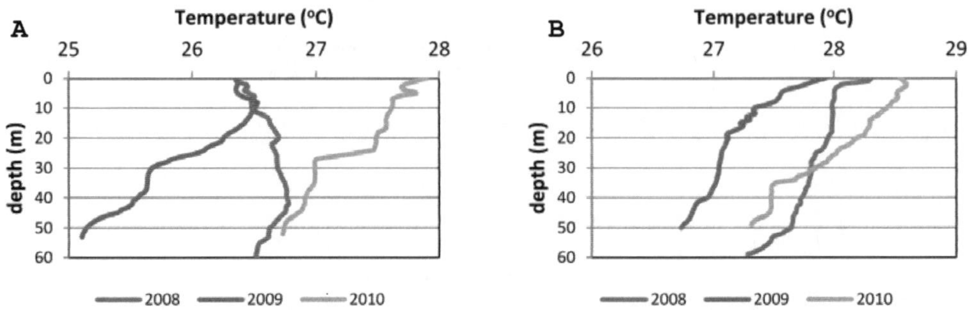

Figure 5. Vertical temperature distribution in station 7 on (a) July and (b) October 2008, 2009, and 2010.

Based on vertical temperature measurement in one of the outer stations in Ambon Bay (Station 7) in July (figure 5a) and October (figure 5b) in the year of 2008, 2009, and 2010, it is apparent that El Niño and La Niña had an impact on vertical temperature compared with normal condition. When El Niño event occurred in 2009, the temperature in AOB became cooler, whereas, during the 2010 La Niña, the temperature was warmer than the 2008 normal condition.

The temperature in Indonesian waters tended to be colder than usual when El Niño occured due to the eastward shift of the warm pool toward the central Pacific, which elevated the ocean temperature there. This condition made the air pressure above the sea surface lower than usual resulted in occurrences of eastward wind anomaly against the westward blowing Trade Winds. This wind anomaly moves water mass in the West Pacific region to the eastern Pacific triggering upwelling in West Pacific region. Water mass with lower temperature and high density moves upward to the surface when upwelling happens, and thus making sea surface temperature in the western Pacific region to be cooler than usual [21].

On the contrary, when La Niña occurs, sea surface temperature in the Niño 3.4 region becomes cooler than usual due to the pile-up of warm water over the Maritime Continent caused by the stronger Trade Winds. This makes the thermocline layer to remain at depth, and the mixed layer could become thicker than usual [22].

The most effective water mass circulation in Ambon Bay happens in the Southeast monsoon when there is upwelling in the Banda Sea [23]. At this time, water mass in thermocline layer is lifted to the surface until it reaches ± 70 m depth which previously was at ± 150m [14]. This water mass is thought to get into the deep AIB through tidal upwelling mechanism replacing the resident water mass because the density is higher than the host density. Observation using 3D hydrodynamic modelling suggests that water mass from the Banda Sea gets into the deep AIB through the narrow sill between June to August [14,15].

Figure 6a shows the vertical temperature distribution in all stations in June 2008 and 2009. The temperature in June 2008 was apparently warmer than in 2009. In June 2008, the water mass circulation in Ambon Bay was not seen and based on the T-S diagram (figure 6b and 6c), it was seen that there was a different water mass characteristic between Ambon Inner and Outer Bay. This condition explains that water mass exchange only happened at the surface and flushing phenomenon was predicted not to occur at this time.

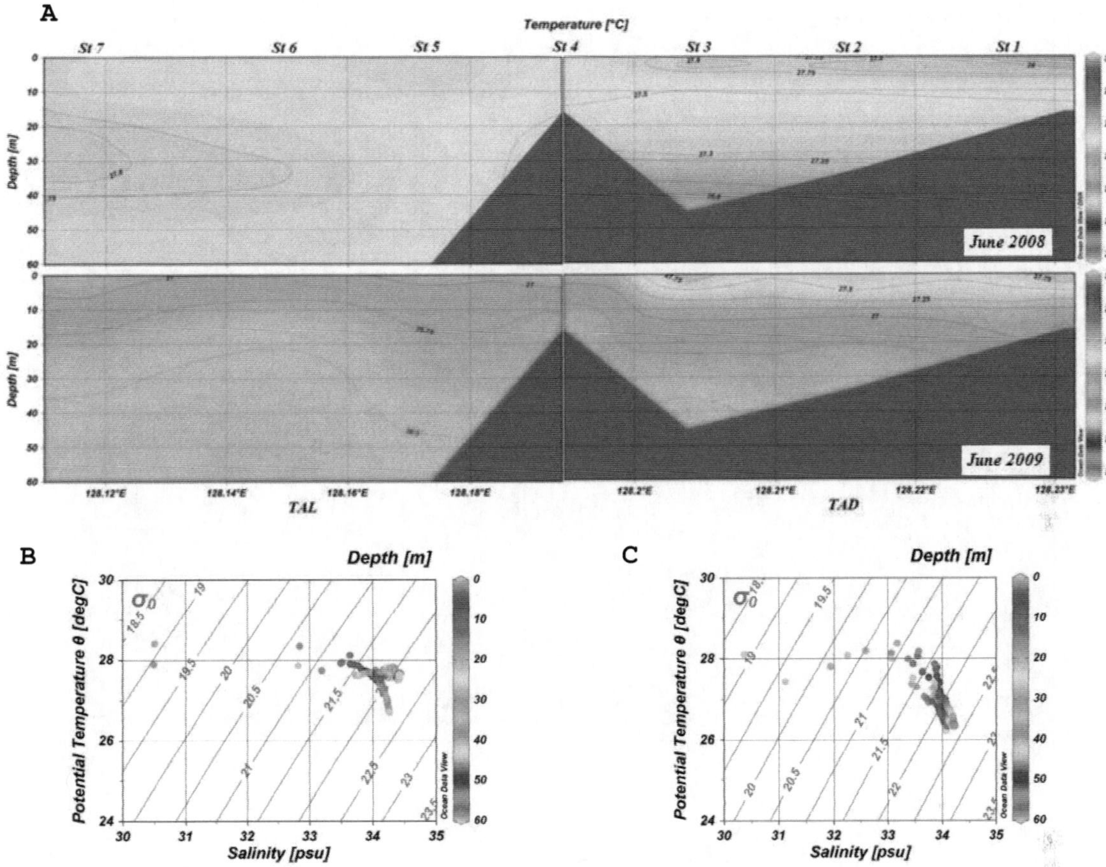

Figure 6. a. Vertical temperature distribution based on locations: Ambon Outer Bay (TAL) and Ambon Inner Bay (TAD) on June 2008 and 2009 and T-S diagrams on (b) June 2008 (c) and 2009.

Observation on July 2008, 2009, and 2010 showed that there was an obvious difference in vertical temperature variability (figure 7). In July 2008, it can be seen that vertical temperature distribution was almost the same in all stations and this observation was shown by the T-S diagrams (figure 7b-d). It was estimated that the water mass from AOB had filled most of AIB.

Both of vertical temperature profiles (July 2009 and July 2010) showed water mass stratification, but there was an obvious difference in the value of temperature distribution (figure 7). The temperature distribution in July 2010 (La Niña) was higher than on July 2009 (El Niño) at which temperature in the surface was still warm even it had entered the Southeast monsoon. In July 2009, vertical temperature profile showed that the appearance of El Niño could strengthen the upwelling process in the Banda Sea where water mass with low temperature (± 25°C) could rise into shallower depth (± 40m). Water mass from the Banda Sea then entered the AOB. It was predicted that flushing phenomenon began to happen at this time.

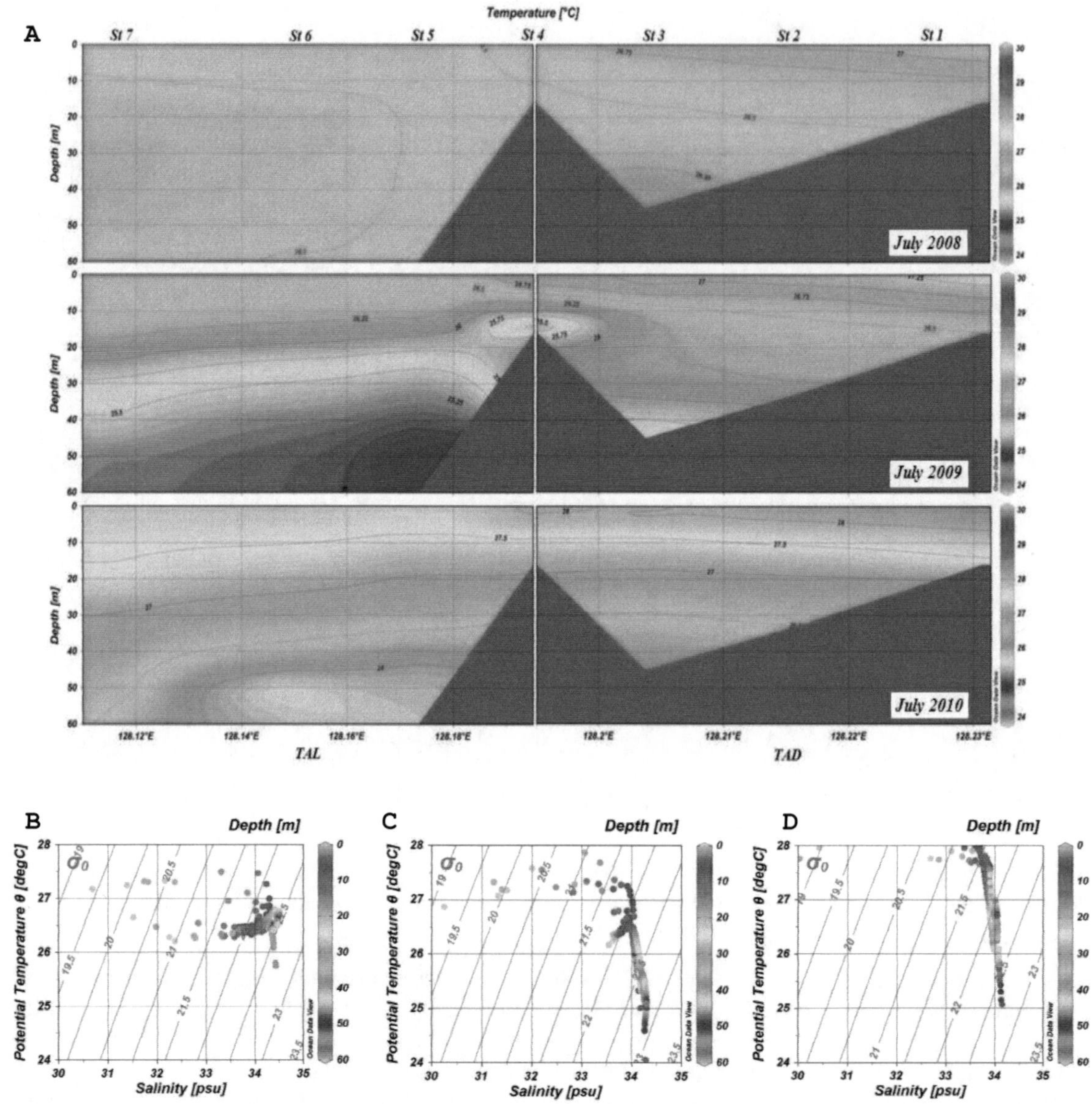

Figure 7. a. Vertical temperature distribution on July 2008, 2009 and 2010 and T-S diagrams in (b) July 2008, (c) 2009, and (d) 2010.

The vertical temperature profile in AOB on September 2009 was warmer than on September 2008 (figure 8a), and flushing phenomenon in Ambon Bay was not seen at this time. This was due to the water mass, which passed through the narrow sill, had a lower density than the water mass in the deep AIB and thus made water mass in the deep AIB trapped and could not exchange with water mass from AOB. We may observe these conditions on the T-S diagram for September 2009 (figure 8b and 8c), where water mass from AOB had a different characteristic to that in AIB.

Based on figure 9, it can be seen that the vertical temperature distributions for October 2008, 2009, and 2010 are different. According to Wyrtki [23], there was a sinking phenomenon where water mass was going down into the bottom of the Banda Sea when entering transition season II until northwest monsoon (October – March). This condition made the water mass density which passes through the bottom of the sill had a lower value than water mass in the deep AIB, and it was estimated that flushing did not happen at this time.

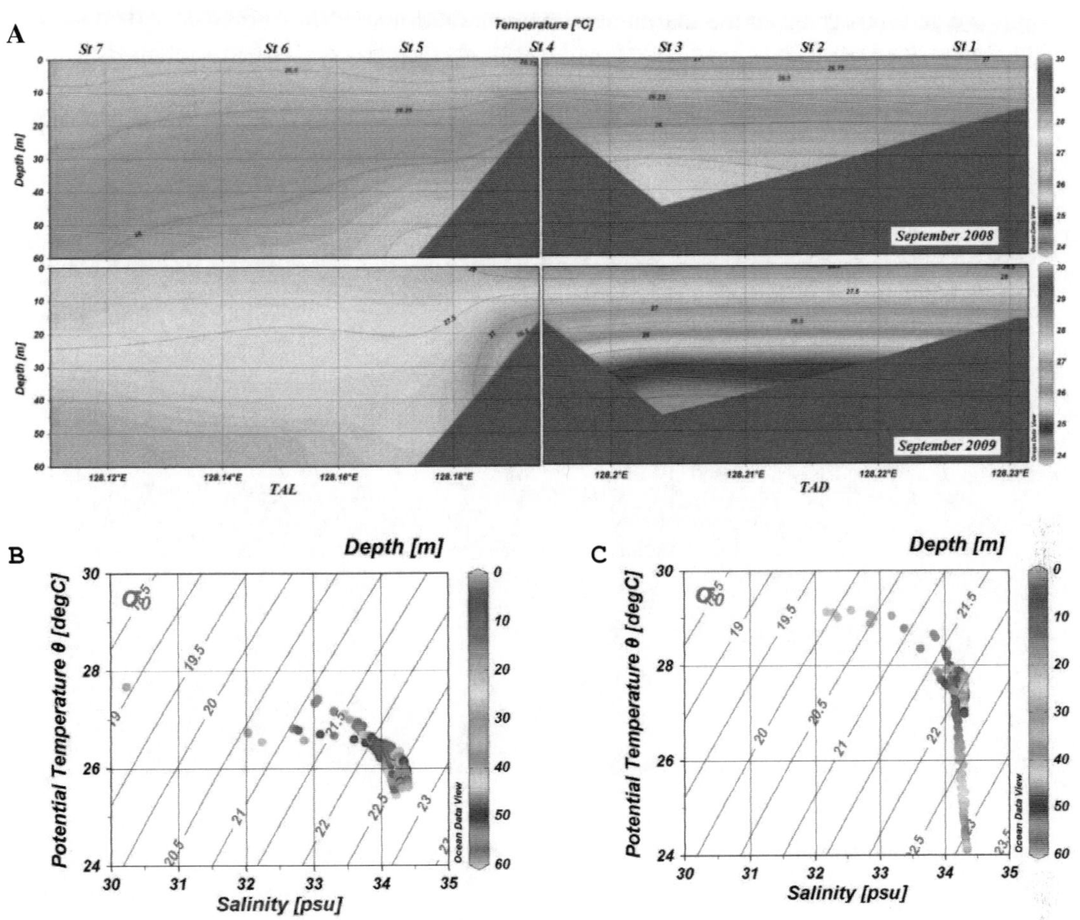

Figure 8.a. Vertical temperature distribution in September in the year of 2008 and 2009 and T-S diagrams in (b) September 2008 and (c) 2009.

It looked as though this condition happened in October 2008 where the temperature in AOB had risen compared with September 2008. This made water mass in AIB trapped and could not go out to AOB, due to water mass density passing through the sill being lower than water mass in the deep AIB (figure 9a). Similar condition also occurred in October 2010, but at this time there was an anomaly in temperature distribution in AOB. At the depth of ± 50m, water mass with lower temperature and high density increased to the surface again and it was estimated that this condition happened because of higher than normal Niño 3.4 SSTs for October 2009. This condition made the temperature in Ambon Bay Waters lower than in September 2009 even when it was entering the transition season II (figure 9b).

The vertical temperature profile in October 2010 showed a higher value than in October 2008 and 2009. This condition did not just happen only in AOB but also in AIB. It is predicted that there was heat diffusion from the surface to the bottom of AIB.

Ambon Bay is a potential fishing area and known as bait fish farm especially in AIB. Most of the fisherman in this area caught small pelagic fish such as *Sardinella* sp., *Rastreliger* sp., *Decapterus* sp., and *Selar* sp. [25]. The changes in the marine environment affect marine organisms that live in it and later affect the fishing activity in this area. Phytoplankton as the basis of life is one of the marine organism that plays an important role in the aquatic productivity. We use the abundance of phytoplankton to assess the water fertility in Ambon Bay. In July 2008, the density of phytoplankton reached 51,701–853,333 cell.m^{-3} [26], this number was higher for July 2009 with phytoplankton density reaching 263,540–5,697,576 cell.m^{-3} [19]. We predict that the higher amount of phytoplankton in July 2009 is related to El Nino which occurred at that time. El Nino triggered flushing in AIB as mentioned above and thus changed the marine environment. Nutrient in the bottom of AIB would be lifted up into the surface when flushing occured, and might enhance productivity in Ambon Bay.

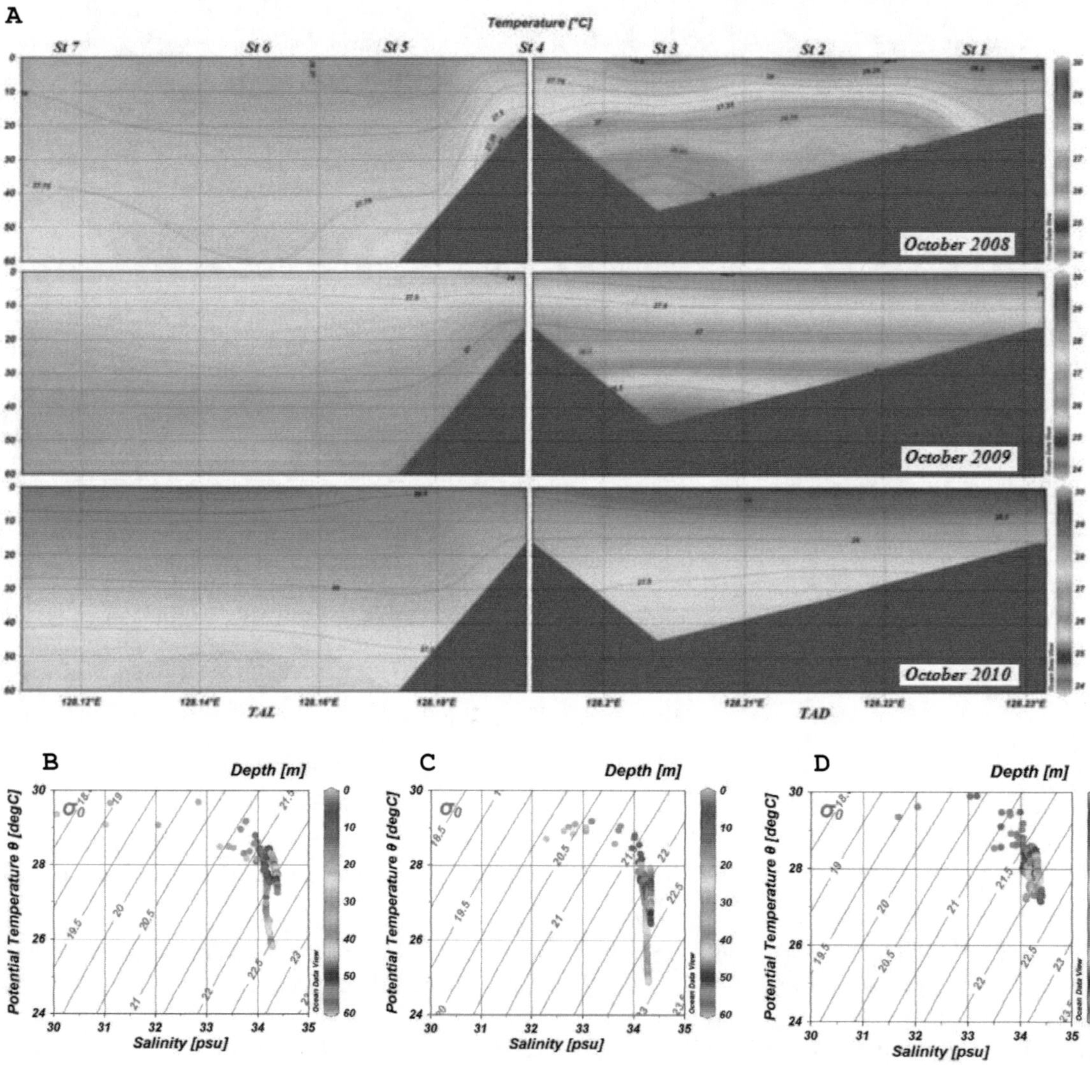

Figure 9. a. Vertical temperature distribution in October 2008, 2009, 2010 and T-S diagrams in October (b) 2008, (c) 2009, and (d) 2010.

4. Conclusion

El Niño and La Niña effect may be observed in temperature variability data in Ambon Bay, and this phenomenon may affect the water mass circulation in Ambon Bay. Water mass circulation in Ambon Bay typically occurs in the southeast monsoon (July). El Niño makes water mass circulation in Ambon Bay to occur earlier, and La Niña makes the circulation to be slower than usual. We still need more data in order to confirm the effect of El Niño and La Niña on water mass circulation in Ambon Bay and how these water mass circulation changes would affect its marine environment.

Acknowledgement

The research activities were funded by DIPA UPT BKBL LIPI Ambon. We thank the anonymous reviewers for their careful reading of our manuscript and their many insightful comments and suggestions. We also thank Abdul Basit for the helpful discussions on ocean current data and analysis.

Reference

[1] Fieux M, Andrié C, Charriaud E, Ilahude A G, Metzl N, Molcard R and Swallow J C 1996 Hydrological and chlorofluoromethane measurements of the Indonesian throughflow entering the Indian Ocean *J. Geophys. Res.* **101**(C5) 12433–54

[2] MacDonald A M 1993 Property fluxes at 30°S and their implications for the Pacific-Indian throughflow and the global heat budget *J. Geophys. Res.* **98**(C4) 6851-68

[3] Kunarso K, Hadi S, Ningsih S N and Baskoro M S 2011 Variabilitas suhu dan klorofil-a di daerah upwelling pada variasi kejadian ENSO dan IOD di perairan Selatan Jawa sampai Timor *Ilmu Kelautan: Indonesian J. Mar. Sci.* **16**(3) 171–180

[4] Safitri M, Cahyarini S Y and Putri M R 2012 Variasi arus arlindo dan parameter oseanografi di Laut Timor sebagai indikasi kejadian ENSO, *Jurnal Ilmu dan Teknologi Kelautan Tropis* **4**(2) 369-377

[5] Hasita F, Zikra M and Suntoyo 2013 Analisa variasi temperatur dan salinitas air laut di perairan Samudra Pasifik akibat pengaruh El Nino dan La Nina *Jurnal Teknik Pomits* **2**(2) G181 – 5

[6] Ward P J, Kummu M and Lall U 2016 Flood frequencies and durations and their response to El Niño Southern Oscillation: Global analysis *J. Hydrol.* **539** 358 – 378.

[7] Trenbert K E, Caron J M, Stepaniak D P and Worley S 2002 Evolution of El Niño–Southern Oscillation and global atmospheric surface temperatures *J. Geophys. Res.* **107**(D8) AAC-5-1-17

[8] Mulyana E 2002, Hubungan antara ENSO dengan variasi curah hujan di Indonesia, *Jurnal Sains & Teknologi Modifikasi Cuaca* **3**(1) 1-4

[9] Rayner N A, Parker D E, Horton E B, Folland C K, Alexander L V, Rowell D P, Kent E C and Kaplan A 2003 Global analyses of sea surface temperature, sea ice, and night marine air temperature since the late nineteenth century *J. Geophys. Res.* **108** (D14) 4407-37

[10] Susanto R D, Gordon A L and Zheng Q 2001 Upwelling along the coast of Java and Sumatera and it's relation to ENSO *J. Geophys. Res. Lett.* **28**(8) 1599 – 1602

[11] Susanto R D and Marra J 2005 Effect of the 1997/98 El Niño on chlorophyll a variability along the southern coasts of Java and Sumatera *Oceanography* **18**(4) 124–7

[12] Basit A, Mudjiono and Putri M R 2008 Monitoring oseanografi fisis Teluk Ambon, *Prosiding Pertemuan Ilmiah Tahunan ISOI* (Bandung: ISOI) pp 41–7

[13] Zijlstra J J, Baars M A, Tijssen S B, Wetsteyn F J, Witte J I J, Ilahude A G and Hadikusumah 1990 Monsoonal effect on the hydrography of the upper waters (<300m) of the eastern Banda Sea and northern Arafura Sea, with special reference to vertical transport processes, *Neth. J. Sea Res.* **25**(4) 431–47

[14] Hamzah M S and Wenno L F 1987 Sirkulasi arus di Teluk Ambon *Laporan Balai Penelitian dan Pengembangan Sumberdaya Laut Pusat Penelitian dan Pengembangan Oseanologi LIPI* (Ambon: BPPSDL P2O LIPI) aaccessed Feb 5 2016 from http://coremap.or.id/downloads/1679.pdf

[15] Anderson J J and Sapulette D 1981 Deep water renewal in Inner Ambon Bay, Ambon, Indonesia *Proc. 4th Int. Coral Reef Symp.* **1** 370–4 (Manila: University of the Philippines) accessed Febr. 10 2016 from http://www.reefbase.org/resource_center/publication/pub_10032 .aspx

[16] Saputra F R T and Lekalette J D 2016 Water mass dynamics in Ambon Bay *Widyariset* **2**(2) 143-152

[17] Schlitzer R 2015 *Ocean Data View* [Internet] accessed Feb 20, 2016 from http://odv.awi.de.

[18] Corvianawatie C, Cahyarini S Y and Putri M R 2016 Reconstruction of sea surface temperature data based on the Sr/Ca of porites coral in Ambon Bay *J. Math. Fund. Sci.* **48**(2) 115–29

[19] UPT BKBL LIPI Ambon 2009 *Research Report* (Ambon: UPT BKBL LIPI Ambon)

[20] National Oceanic and Atmospheric Administration 2017 *Niño34* [Internet] accessed on Apr 25 2016 from https://www.esrl.noaa.gov/psd/gcos_wgsp/Timeseries/Data/Niño34.long.anom.data

[21] Neelin J D and Mojib L 1998 El Niño dynamics *Physics Today* **51**(12) 32-36

[22] Suplee C 1999 El Niño / La Niña: Nature's vicious cycle *Nat. Geogr. Mag.* **195**(3) 72-95

[23] Wyrtki K 1961 Physical Oceanography of the Southeast Asian Waters *Naga Rep. 2* pp 1-195
[24] Corvianawatie C 2014 *Mekanisme Pertukaran Massa Air di Teluk Ambon Berdasarkan Model Asimilasi Densitas* Master Thesis (Bandung: Institut Teknologi Bandung) 37 pp
[25] Syahailatua A 1999 Komunitas fauna ikan yang tertangkap dengan jaring pantai dan bagan di Ambon *Oseanologi Indonesia* **31** 41-55
[26] UPT BKBL LIPI Ambon 2009 *Research Report* (Ambon: UPT BKBL LIPI Ambon)

Institute of Physics
Dirac House, Temple Back
Bristol BS1 6BE UK

ISSN: 1755-1307
ISBN 978-1-5108-7010-9

CURRAN ASSOCIATES INC.
proceedings.com

30th Workshop on Recent Developments in Computer Simulation Studies in Condensed Matter Physics 2017

Journal of Physics: Conference Series Volume 921

Athens, Georgia, USA
20 – 24 February 2017

IOP
Institute of Physics